COSMOLOGICAL
RELATIVITY

The Special and General Theories
for the Structure of the Universe

COSMOLOGICAL RELATIVITY

The Special and General Theories for the Structure of the Universe

Moshe Carmeli
Ben Gurion University, Israel

World Scientific

NEW JERSEY · LONDON · SINGAPORE · BEIJING · SHANGHAI · HONG KONG · TAIPEI · CHENNAI

Published by

World Scientific Publishing Co. Pte. Ltd.

5 Toh Tuck Link, Singapore 596224

USA office: 27 Warren Street, Suite 401-402, Hackensack, NJ 07601

UK office: 57 Shelton Street, Covent Garden, London WC2H 9HE

British Library Cataloguing-in-Publication Data
A catalogue record for this book is available from the British Library.

COSMOLOGICAL RELATIVITY
The Special and General Theories for the Structure of the Universe

ISBN-13 978-981-270-075-9
ISBN-10 981-270-075-7

Printed in Singapore

PREFACE

This book is written for those who are interested in the large-scale structure of the Universe, whether they are relativists, cosmologists, astronomers or just physicists. It summarizes the research and results of the cosmological special relativity as well as the cosmological general relativity. The cosmological special relativity is presented along the lines of Einstein's special relativity and it is needed even though we know that the cosmos is actually filled up with gravity.

Its generalization to a full theory of gravitation, cosmological general relativity, in four dimensions of space and velocity is then followed up. These theories make new predictions that are not included in Einstein's special and general theories of relativity. This fact is not an indication that anything is wrong with Einstein's theories. On the contrary, Einstein's theories have been very successful in cosmology and in local (Lorentz) invariance that have led physics to one of its greatest achievements.

Cosmological general relativity is then extended to five dimensions so as to have spacetime and velocity. Cosmological general relativity does not have, neither it needs, a cosmological constant. It was also shown by Dr John G. Hartnett that there is no need to assume the existence of dark matter in the Universe (Intern. J. Theor. Phys. **44** (3), 359–372 and **44** (4), 485–492 (2005)). It also means there is no need to the assumption of dark matter in the halos of rotating and spiral galaxies. Among other results, the theory shows that the Universe is expanding with an acceleration with a positive pressure (and not negative as in other theories). Likewise, the Big Bang time in this theory is equal to 13.56 ± 0.48 Gyr, in agreement with the recent results obtained by NASA's WMAP according to which initial cosmic inflation occurred 13.7 ± 0.2 Gyr ago.

The book is written in a clear and pedagogical way so as to enable the non-experts as well as experts to easily understand its content. We have tried to use as few formulas as possible to the extent that the text will still be accurate.

It is a pleasure to thank Dr Julia Goldbaum for her great help, in both the scientific and the technical contents, in writing this book. It is also a great pleasure to thank Dr John G. Hartnett, University of Western Australia, and Dr Firmin J. Oliveira, Institute of Space Research, Hawaii, for their great contributions to the understanding and the expanding of the theories presented in this book.

Beer Sheva, Israel   *Moshe Carmeli*
October, 2006

CONTENTS

CHAPTER 1

INTRODUCTION

1.1 Remarks on cosmological special relativity

A few remarks on cosmological special relativity are worth to be mentioned now.

First we have the principle of cosmological relativity according to which the laws of physics are the same at all cosmic times. This is an extension of Einstein's principle of relativity according to which the laws of physics are the same in all coordinate systems moving with constant velocities. In cosmology the concept of time $(t = x/v)$ replaces that of velocity $(v = x/t)$ in Einstein's special relativity.

Second, we have the principle that the Big Bang time τ is always constant with the same numerical value, no matter at what cosmic time it is measured. This is obviously comparable to the assumption of the constancy of the speed of light c in Einstein's special relativity.

Velocity in an expanding Universe is not absolute just as time is not absolute in special relativity. Velocity now depends on at what time an object (or a person) is located; the more backward in time, the slower velocity progresses, the more distances contract, and the heavier the object becomes. In the limit that the object approaches the Big Bang event, velocities and distances contract to nothing, and the object's energy becomes infinite.

In Einstein's special relativity things depend on the velocity: The faster the object moves, the slower time progresses, the more distances contract, and the heavier the object becomes.

These features are discussed in details in the next chapter.

CHAPTER 2

COSMOLOGICAL SPECIAL RELATIVITY

In this Chapter we present cosmological special relativity. This is done in analogous to Einstein's special relativity of space and time. In our case it will be a four-dimensional space and velocity (later on will be extended to five dimensions by including the time). These are exactly the variables that astronomers measure, the distances to galaxies and their receding velocities (the redshifts). But first we start with Einstein's special relativity.

2.1 The Galileo transformation and its generalization

We start with Einstein's special relativity and how it came about.

2.1.1 The Galileo transformation

One starts with the Galileo transformation that connects coordinate systems moving with constant velocities with respect to each other. Such a transformation between two coordinate systems is given by (see Fig. 2.1)

$$x' = x + Vt, \quad t' = t, \tag{2.1}$$

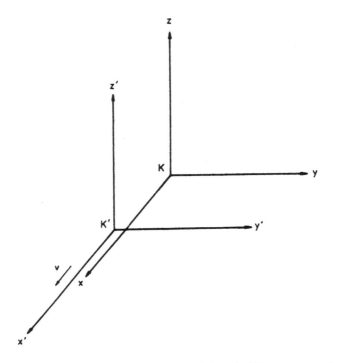

Figure 2.1: Two coordinate systems K and K', one moving with respect to the other with a velocity v in the x-direction.

where the axes y and z were kept unchanged, and V is constant. This is the well-known Galileo transformation. If we assume that x and x' are the coordinates of a particle in the two systems, and taking now the time derivative of the above equation, we obtain

$$v' = v + V, \qquad (2.2)$$

where v' and v are the velocities of the particle in the two systems of coordinates. As we see the velocity in the new coordinate system is equal to that in the old coordinate system with the addition of the constant velocity V. This is of course an expected result. There is no problem here. The velocities are added linearly.

2.1.2 Difficulties with light

But what happens if x and x' are assumed to be the coordinates of a light pulse. Instead of Eq. (2.2) we now obtain

$$c' = c + V, \tag{2.3}$$

where c is the speed of light in the old system and c' is the speed in the new system. Thus we see that the speed of light is now larger than c. But this is impossible. It violates the assumption that the speed of light is constant with respect to all moving coordinate systems (one of the two postulates of Einstein's special relativity). Hence, the Galileo transformation is not adequate for light. It has to be generalized to accommodate the weird behavior of light. The generalization of the Galileo transformation then leads to the Lorentz transformation that connects space and time.

2.1.3 Role of velocity in classical physics

The role of the velocity in classical physics is not only important but also intrigue. Consider, for example, the Lagrangian of a free particle: $L = a\,\mathbf{R}\cdot\mathbf{R} + b\,\mathbf{R}\cdot\dot{\mathbf{R}} + c\,\dot{\mathbf{R}}\cdot\dot{\mathbf{R}}$, where a, b and c are some constants, $\mathbf{R} = (x, y, z)$ are the Cartesian coordinates of the particle and the overdot denotes a time derivative. The first two terms depend on the location of the particle and hence they cannot represent a physical situation that gives a description of the particle. Only the third term can represent the particle. Using the Lagrange equation, then $(\partial/\partial\dot{x})L$ gives $2c\dot{x}$. Consequently the Lagrange equation gives $2c\ddot{x} = 0$, and the same for the coordinates y and z. This was for a free particle. Accordingly, the velocity is constant. Thus when there are no external forces the particle will move with a constant velocity. This is the first of Newton's three laws of motion.

In the next section the analogue to the Galileo transformation in cosmology will be presented.

2.2 Nonrelativistic cosmological transformation

In cosmology, the same procedure repeats itself but now for the expanding Universe.

2.2.1 Nonrelativistic transformation

The nonrelativistic transformation between two cosmic systems of coordinates is given by

$$x' = x + Tv, \quad v' = v, \tag{2.4}$$

where v is a parameter which has the dimension of velocity and T is the relative cosmic time between the two coordinate systems. Taking now the derivative of the above equation with respect to v we obtain

$$\frac{dx'}{dv} = \frac{dx}{dv} + T.$$

But $dx'/dv = t'$ and $dx/dv = t$, where t' and t are the cosmic times at the two coordinate systems. Thus (see Fig. 2.2)

$$t' = t + T. \tag{2.5}$$

2.2.2 Difficulties at the Big Bang

In cosmology, observers are supposed to be in different cosmic times rather than moving with respect to each other at constant velocities. Equation (2.5) is the analogue of Eq. (2.2) in Einstein's special relativity. There is no problem here also for finite times t and t'. But what happens if we take t to be equal to the Big Bang time (which is now assumed, in analogy to Einstein's special relativity, to be constant at all cosmic times). We obtain

$$\tau' = \tau + T, \tag{2.6}$$

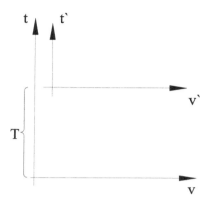

Figure 2.2: Two cosmic frames K and K', time interval between them being T.

where τ denotes the Big Bang time (τ is equal to 13.56±0.48Gyr) and τ' is the new Big Bang time. Thus the cosmic time now is larger than the Big Bang time and that is impossible by our assumption. Therefore the above nonrelativistic transformation should be replaced by a larger transformation that involves both space and velocity. This is similar to the situation in Einstein's special relativity. In the previous case that leads to the Lorentz transformation and in cosmology it leads to the cosmological transformation.

In cosmology as we have seen, we have to deal with quantities that depend on the velocity rather than on time. If R represents the location of a galaxy with respect to an observer and v is the receding velocity of the galaxy then one can, in analogue to what was done in the preceding section write the Lagrangian: $L = (dR/dv)^2$, neglecting terms that depend on the location. From the Lagrange equation one then has $(\partial/\partial(dR/dv))L = 2(dR/dv)$. And the Lagrange equation gives $d^2R/dv^2 = 0$. Thus $dR/dv = \text{const}$, and $R = \text{const} \times v$. One then easily identify the constant with the reciprocal of the Hubble parameter H_0 for the case of negligible gravity (flat space).

In the next two sections the Lorentz transformation and its extension to cosmology are presented.

2.3 Extension to the Lorentz transformation

The easiest way to obtain the generalization of the nonrelativistic Galileo transformation is as follows.

2.3.1 Invariance of light propagation

Since light is supposed to be constant with respect to all inertial systems, one can write

$$c^2 t^2 - (x^2 + y^2 + z^2) = c^2 t'^2 - (x'^2 + y'^2 + z'^2). \qquad (2.7)$$

The above relation is a generalization to four dimensions of the familiar three-dimensional rotation of coordinates in the Euclidean space. However, now the metric has the signature $(+ - --)$ rather than with just $(+ + +)$ in the nonrelativistic case.

2.3.2 The Lorentz transformation

A direct result of the invariance relationship (2.7) is

$$x' = \frac{x - (v/c)\,ct}{\sqrt{1 - v^2/c^2}}, \quad ct' = \frac{ct - (v/c)\,x}{\sqrt{1 - v^2/c^2}}, \qquad (2.8)$$

where $y' = y$ and $z' = z$, and $0 \le v < c$. Equations (2.8) are the familiar Lorentz transformation which are the basis of most our theories in local physics.

In the next section we derive the cosmological transformation, the analogue of the Lorentz transformation for cosmology.

2.4   Extension to the cosmological transformation

2.4.1   Invariance of the Big Bang time

In the cosmological case the analogue of Eq. (2.7) is

$$\tau^2 v^2 - (x^2 + y^2 + z^2) = \tau^2 v'^2 - (x'^2 + y'^2 + z'^2), \qquad (2.9)$$

where τ is the Big Bang time. The above equation expresses the invariance of the Big Bang time at all cosmic times.

2.4.2   The cosmological transformation

It yields

$$x' = \frac{x - (t/\tau)\,\tau v}{\sqrt{1 - t^2/\tau^2}}, \quad \tau v' = \frac{\tau v - (t/\tau)\,x}{\sqrt{1 - t^2/\tau^2}}, \qquad (2.10)$$

where $y' = y$ and $z' = z$. In the above equations t is the cosmic time measured backward with respect to us (it is zero now and equals to τ at the Big Bang). In this way the Lorentz and the cosmological transformations are seen to be similar in their behavior, one with respect to space and time and the other with respect to space and velocity. When $v \to c$ we have all the "difficulties" of light. So is the situation with the cosmological transformation when $t \to \tau$ we have the "troubles" of the Universe close to the Big Bang.

The cosmological transformation predicts very interesting physical phenomena of the contraction of velocities and distances similar to those of the Lorentz transformation. It leads to new physics in cosmological invariance just as the Lorentz transformation in local invariance. There is no escape from cosmological invariance just as there was no escape from local invariance.

In the next section we find the temperature of the Universe in the approximation when gravity is neglected (flat space) and compare it to that when matter is presented.

2.5   Temperature of the Universe

Denote the temperature of the Universe at a cosmic time t by T and that at present by T_0 (=2.73K), we then have

$$T = \frac{T_0}{\sqrt{1 - t^2/\tau^2}}, \qquad (2.11)$$

where t is the cosmic time measured with respect to us now and $\tau = 13.56 \pm 0.48$Gyr. For temperatures T at very early times, we can use the approximation $t \approx \tau$, thus

$$1 - t^2/\tau^2 = (1 + t/\tau)(1 - t/\tau) \approx 2(1 - t/\tau)$$
$$= (2/\tau)(\tau - t) = 2\tilde{t}/\tau, \qquad (2.12)$$

where $\tilde{t}$ is the cosmic time with respect to the Big Bang. Using this result we obtain

$$T = T_0 (\tau/2)^{1/2} \tilde{t}^{-1/2}. \qquad (2.13)$$

The thermodynamical formula that relates the temperature to the cosmic time with respect to the Big Bang is well known and given by

$$T = \left(\frac{45\hbar^3}{32\pi^3 k^4 G}\right)^{1/4} \tilde{t}^{-1/2}, \qquad (2.14)$$

where k is Boltzmann's constant, G is Newton's gravitational constant, $\hbar = h/2\pi$ and h is Planck's constant. As is seen, both equations show that the temperature T depends on $\tilde{t}^{-1/2}$. The coefficients appearing before the $\tilde{t}^{-1/2}$, however, are not identical. A simple calculation shows

$$\left(\frac{45\hbar^3}{32\pi^3 k^4 G}\right)^{1/4} = 1.52 \times 10^{10} \text{Ks}^{1/2}, \qquad (2.15)$$

and
$$T_0 \left(\tau/2\right)^{1/2} = 1.21 \times 10^9 \text{Ks}^{1/2}. \tag{2.16}$$

In the above we have used $\hbar = 1.05 \times 10^{-34}$Js, $k = 1.38 \times 10^{-23}$J/K, $G = 6.67 \times 10^{-11} \text{m}^3/\text{s}^2\text{Kg}$, $T_0 = 2.73$K, $\tau = 13.56$Gyr. Accordingly we can write for the temperatures in both cases

$$T \approx 1.5 \times 10^{10} \text{Ks}^{1/2} \tilde{t}^{-1/2}, \tag{2.17}$$

and
$$T \approx 1.2 \times 10^9 \text{Ks}^{1/2} \tilde{t}^{-1/2}. \tag{2.18}$$

The ratio between them is approximately 13.

2.5.1 Conclusions

It thus appears that the dominant part of the plasma energy of the early Universe has gone to the creation of matter appearing now in the Universe, and only a small fraction of it was left for the background cosmic radiation.

In the next section we give the line elements in both Einstein's and cosmological special relativity.

2.6 Line elements in Einstein's special relativity and in cosmological special relativity

2.6.1 Line element in Einstein's relativity

In Einstein's special relativity the line element is given by

$$ds^2 = c^2 dt^2 - (dx^2 + dy^2 + dz^2). \tag{2.19}$$

ds describes the infinitesimal distance between two adjoint points in the Minkowskian space and time. The quantity ds/c is often called the proper time. It describes the time as measured by a local clock.

2.6.2   Light propagation

For light propagation the null condition $ds = 0$ should be added to Eq.(2.19) and thus

$$dx^2 + dy^2 + dz^2 = c^2 dt^2. \qquad (2.20)$$

Hence the proper time of light is zero. The propagation of light in empty space is usually described by the light cone (see Fig. 2.3).

2.6.3   Line element in cosmological special relativity

In cosmological special relativity the line element is

$$ds^2 = \tau^2 dv^2 - (dx^2 + dy^2 + dz^2), \qquad (2.21)$$

where ds now describes the infinitesimal distance between two adjoint points in the Minkowskian space and velocity, and the quantity ds/τ might be called the proper velocity. It gives a measure of the velocity for a local observer.

2.6.4   Hubble's expansion

For the Hubble expansion with negligible gravity one adds the null condition $ds = 0$, and the Universe expands according to

$$dx^2 + dy^2 + dz^2 = \tau^2 dv^2. \qquad (2.22)$$

The proper velocity in this case is zero. Galaxies can then in this case be presented in what is called the galaxy cone similarly to the light cone for light (see Fig. 2.4).

In the next section we give a very simple derivation for the inflation of the Universe at the early times.

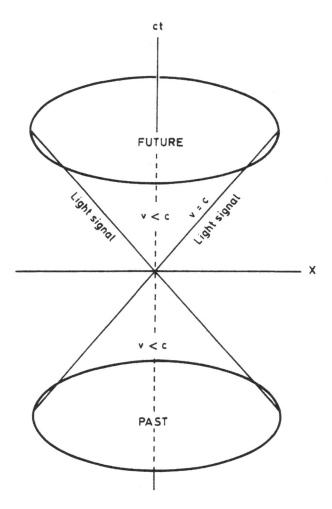

Figure 2.3: The light cone in two dimensions, $x^0(= ct)$ and $x^1(= x)$. The propagation of two light signals in opposite directions passing through $x = 0$ at time $t = 0$, is represented by the two diagonal straight lines. The motion of finite-mass particles, on the other hand, are represented by straight lines in the interior of the light cone. (Compare the galaxy cone given in Fig. 2.4.)

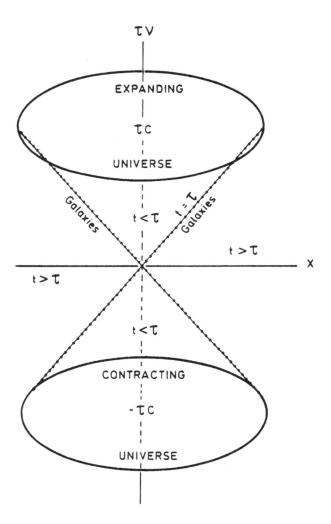

Figure 2.4: The galaxy cone in cosmological relativity, describing the cone in the x - v space satisfying $x^2 - \tau^2 v^2 = 0$ represents the three-dimensional space. The heavy dots describe galaxies. The galaxy cone represents the locations of the galaxies at a given time rather than their path of motion in the real space. (Compare the light cone given in Fig. 2.3.)

2.7 Inflation of the Universe

Much has been written about inflation in the early Universe and in different varieties associated with antigravity or scalar fields. In cosmological special relativity inflation at the early Universe comes out very naturally and can be shown as follows.

Dividing Eq. (2.21) by ds^2 we obtain

$$\tau^2 \left(\frac{dv}{ds}\right)^2 - \left[\left(\frac{dx}{dv}\right)^2 + \left(\frac{dy}{dv}\right)^2 + \left(\frac{dz}{dv}\right)^2\right] \left(\frac{dv}{ds}\right)^2$$

$$= \left(\tau^2 - t^2\right) \left(\frac{dv}{ds}\right)^2 = 1. \tag{2.23}$$

2.7.1 Matter density

Multiplying now this equation by ρ_0^2, the matter density of the Universe at the present time, we obtain for the matter density at a past time t

$$\rho = \tau \rho_0 \frac{dv}{ds} = \frac{\rho_0}{\sqrt{1 - \dfrac{t^2}{\tau^2}}}. \tag{2.24}$$

Since the volume of the Universe is inversely proportional to its density, it follows that the ratio of the volumes at two backward cosmic times t_1 and t_2 with respect to us is given by $(t_2 < t_1)$,

$$\frac{V_2}{V_1} = \sqrt{\frac{1 - t_2^2/\tau^2}{1 - t_1^2/\tau^2}} = \sqrt{\frac{(\tau - t_2)(\tau + t_2)}{(\tau - t_1)(\tau + t_1)}}. \tag{2.25}$$

2.7.2   Ratio of volumes

For times t_1 and t_2 very close to τ we can assume that $\tau + t_2 \approx \tau + t_1 \approx 2\tau$. Hence

$$\frac{V_2}{V_1} = \sqrt{\frac{T_2}{T_1}}, \tag{2.26}$$

where $T_1 = \tau - t_1$ and $T_2 = \tau - t_2$. Thus T_1 and T_2 are the cosmic times as measured with respect to the Big Bang event. For $T_2 - T_1 \approx 10^{-23}$s, and $T_2 \ll 1$s, we thus obtain

$$\frac{V_2}{V_1} = \sqrt{\frac{10^{-32}}{T_1}} = \frac{10^{-16}}{\sqrt{T_1}}. \tag{2.27}$$

For $T_1 \approx 10^{-132}$s, for example, we obtain $V_2 \approx 10^{50}V_1$.

The above result conforms with inflationary Universe theory without assuming any model (such as the Universe is propelled by a sort of antigravity), as the recent NASA's WMAP shows.

In the next section we will find that the Universe has a minimal acceleration.

2.8   Minimal acceleration in Nature

We have

$$t = \frac{x}{v} = \frac{dx}{dv} = \frac{v}{a}, \tag{2.28}$$

where a is the acceleration. Therefore

$$t_{max} = \tau = \left[\frac{v}{a}\right]_{max} = \frac{c}{a_{min}}. \tag{2.29}$$

It thus follows that in Nature there is a minimal acceleration associated with the expansion of the Universe. This result is due to the structure of the Universe and it is not associated with gravity. It is given by

$$a_{min} = \frac{c}{\tau} = \frac{3 \times 10^{10} cm/s}{4.28 \times 10^{17} s} = 7.01 \times 10^{-8} cm/s^2. \tag{2.30}$$

2.8.1   Relation to Pioneer spacecrafts

This value for the minimal acceleration gives answer to the anomalous acceleration of $(8.74 \pm 1.33) \times 10^{-8} \mathrm{cm/s^2}$ discovered in the long journeys of the satellites Pioneer 10 and Pioneer 11 sent by NASA.

In the next section we derive a formula for the redshift in the Universe similar to what is known as the Doppler shift for light.

2.9   Redshift and cosmic time

Problems with redshift and cosmic time are of considerable importance in cosmology. In particular one would like to know if there is any direct relationship between these two physical quantities. We know that in electrodynamics when a charged particle moves with acceleration and radiates electromagnetic waves there is a redshift when the particle is moving away from the observer. This is known as the Doppler shift. Is there any similar relationship in cosmology?

In this section we derive a formula

$$T = \frac{2h^{-1}}{1 + (1 + z)^2},\tag{2.31}$$

where h is the Hubble parameter for the Universe with negligible gravity, and might be taken as $h = 72.17 \pm 0.84 \mathrm{km/s\text{-}Mpc}$, and z is the redshift. The time T is now measured from the Big Bang onward, thus $T = \tau - t$, where t is measured backward. The formula is valid for all z.

Th Universe expands, of course, by the Hubble law $x = H_0^{-1} v$, where H_0 is the Hubble parameter at the present time. But one cannot use this law directly to obtain a relation between z and t. So we start by assuming that the Universe is empty of gravity. As we have seen, one can describe the property of expansion as a null-vector in the flat four dimensions of space and the expanding velocity v.

As was shown in Section 2.6, the cosmological line element was given by Eq. (2.21), where τ is the Big Bang time , the reciprocal of the Hubble parameter H_0 in the limit of zero gravity, and it is a constant in this epoch of time. When $ds = 0$ one gets the Hubble expansion with no gravity.

Space and time coordinates transform according to the Lorentz transformation given by Eq. (2.8) in ordinary physics. In cosmology the coordinates transform by the cosmological transformation given by Eq. (2.10), where t is the cosmic time with respect to us now.

Comparing the Lorentz transformation and the cosmological transformation shows that the cosmological one can formally be obtained from the Lorentz transformation by changing v to t and c to τ $(v/c \rightarrow t/\tau)$. Thus the transfer from ordinary physics to the expanding Universe, under the above assumption of empty space, for null four-vectors is simply achieved by replacing v/c to t/τ, where t is the cosmic time measured with respect to us now.

We now use the above description as follows. In electrodynamics the electromagnetic radiation is described by its frequency ω and the wave vector $\mathbf{k}$. Using the wave four-vector $(\omega, \mathbf{k})$ one can easily derive the transformation of ω and $\mathbf{k}$ from one coordinated system to another. This then gives the Doppler effect.

A charged particle receding from the observer with a velocity v and emitting electromagnetic waves will experience a frequency shift given by

$$\omega = \omega' \sqrt{\frac{1 + v/c}{1 - v/c}}, \tag{2.32}$$

where ω' and ω are the frequencies of the emitted radiation received from the particle at velocity v and at rest, respectively.

And thus a redshift is obtained from

$$1 + z = \sqrt{\frac{1 + v/c}{1 - v/c}}. \tag{2.33}$$

In our case τ replaces c and t replaces v (v/c goes over to t/τ), thus getting

$$1 + z = \sqrt{\frac{1 + t/\tau}{1 - t/\tau}}. \tag{2.34}$$

Rearranging, we get

$$\frac{t}{\tau} = \frac{(1 + z)^2 - 1}{(1 + z)^2 + 1}. \tag{2.35}$$

By using $\tau = 1/h$, and $T = \tau - t$, where now T is measured with respect to the Big Bang time, in the above formula, one easily obtains Eq. (2.31) (see Fig. 2.5 for the case of backward time and Fig. 2.6 for forward time). The two figures give the dependence of the redshift z on the cosmic time, in the two options of forward time T beginning with the Big Bang and the backward cosmic time t starting now.

In this section we have derived a simple formula valid for the case of negligible gravity and all redshift values in the Universe. The formula relates the cosmic time t since the Big Bang, for an earth observer at present epoch, to the measured redshift z of light emitted at time t. The formula could be useful for identifying objects at the early Universe since we can go back in time as far as we desire but not to the Big Bang event itself at which the redshift becomes infinity.

In the next section we present field equations that depend on the redshift (velocity) rather than on time.

2.10 Field equations of a different kind

In 1939 Eugene P. Wigner has found all irreducible representations of the nonhomogeneous Lorentz group and the field

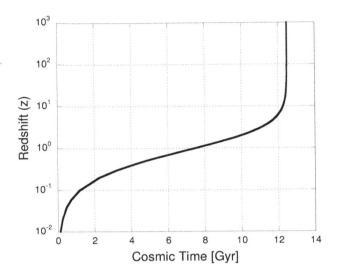

Figure 2.5: The redshift z vs. the backward cosmic time, at which the time now is zero and it is 13.56Gyr at the Big Bang.

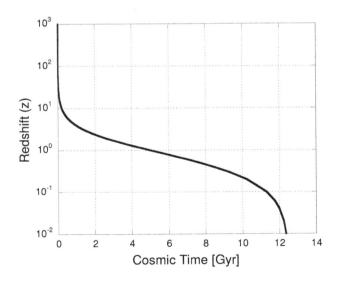

Figure 2.6: The redshift z vs. the forward cosmic time, at which the time at the Big Bang is zero and now it is 13.56Gyr.

equations associated with each representation. The representations were designed by the spin and mass of particles. He found that all standard field equations, such as the Dirac equation, was part of that scheme. Does that mean there are no other field equations? In the following we give two examples of a new kind of field equations in the four dimensions of space and velocity.

2.10.1 Examples

Example 1: A scalar particle with non-zero mass,

$$\hbar^2 \left(\frac{1}{\tau^2} \frac{\partial^2}{\partial v^2} - \nabla^2 \right) \Psi = M^2 c^2 \Psi, \tag{2.36}$$

where $\hbar$ is Planck's constant divided by 2π. This is actually the familiar Klein-Gordon equation but now it is written in the four dimensions of space and velocity, with no time-dependence. Thus retardation in time of the solution will not appear in the form of $t - r/c$ but now is replaced by a delay in velocity, $v - r/\tau$. It is a Higgs scalar.

Example 2: Vector field that is similar to electrodynamics in spacetime. The field will be denoted as the skew-symmetric tensor $f_{\mu\nu}$. The dual to this field is given by $^\star f^{\alpha\beta} = \frac{1}{2} \epsilon^{\alpha\beta\mu\nu} f_{\mu\nu}$ and satisfies $\partial^\star f^{\alpha\beta} / \partial x^\beta = 0$ where ϵ is the totally antisymmetric tensor. The field equations are identical to Maxwell's equations but now in the four dimensions of space and velocity,

$$\frac{\partial f^{\alpha\beta}}{\partial x^\beta} = \frac{4\pi}{\tau} j^\alpha, \tag{2.37}$$

where j^α is the current. No mention of time in the above equations since everything is now measured at an instant of time.

We can also represent $f_{\mu\nu}$ as a rotor of a fourvector A_μ,

$$f_{\mu\nu} = \frac{\partial A_\mu}{\partial x^\nu} - \frac{\partial A_\nu}{\partial x^\mu}, \tag{2.38}$$

with the possibility of changing A_μ by a term of the form $\partial\phi/\partial x^\mu$ without affecting $f_{\mu\nu}$, thus $\partial A^\alpha/\partial x^\alpha = 0$. The field equations then yield

$$\left(\frac{1}{\tau^2}\frac{\partial^2}{\partial v^2} - \nabla^2\right)A_\alpha = -j_\alpha. \tag{2.39}$$

This is the familiar wave equation but now it is written in the four dimensions of space and velocity (redshift). The change is not just mathematical, the physics is now different. The retardation in time of solutions of the above equation is replaced by the velocity delay $v - r/\tau$.

The particle described above is a neutral Higgs vector and it does not emit electromagnetic radiation since the wave equation (2.39) does not involve the usual time derivatives. Instead, it radiates what might be called *cosmic waves* in the space of three dimensions and velocity. It thus appears that the Higgs particle is associated with the velocity of the expansion of the Universe. The properties of these waves should be further studied.

We thus see that cosmology opens a new avenue in physics that has not been revealed previously.

In the next section the cosmological special relativity is extended to five dimensions of space, time and velocity (redshift).

2.11   Cosmological special relativity in five dimensions

We are now in a position to extend cosmological special relativity to five dimensions by adding the time to the four dimensions of space and velocity. Accordingly, the coordinates will be taken as $x^\mu = (x^0, x^1, x^2, x^3, x^4) = (ct, x, y, z, \tau v)$. Thus Greek letters take the values $0,\ldots,4$. Notice that the time coordinate is now taken as the first and the velocity coordinate as the last one, and in the rest of this section we choose units in which $c = \tau = 1$.

The line element is now given by

$$ds^2 = dt^2 - (dx^2 + dy^2 + dz^2) + dv^2, \qquad (2.40)$$

which can be written in the simple form $ds^2 = \eta_{\mu\nu}dx^\mu dx^\nu$, where $\eta_{\mu\nu}$ is a generalized Minkowskian metric in five dimensions, given by (1,-1,-1,-1,1), with signature -1.

The invariance in these five dimensions is written as

$$dt'^2 - (dx'^2 + dy'^2 + dz'^2) + dv'^2 = dt^2 - (dx^2 + dy^2 + dz^2) + dv^2. \qquad (2.41)$$

2.11.1 Subtransformations

A subtransformation is obtained if one takes v to be unchanged. We then have the four-dimensional Lorentz transformation $(t, x, y, z) \rightarrow (t', x', y', z')$. This is suitable for observers located at inertial frames moving with constant velocities. The parameter of the transformation between two frames is V/c ($c = 1$), where V is the relative velocity. The transformation of the coordinates is the familiar Lorentz transformation given previously in Section 2.3.

For invariant t we get the cosmological transformation $(x, y, z, v) \rightarrow (x', y', z', v')$. An observer here, located in a cosmic frame makes observations at a fixed time. The parameter of the transformation between two cosmic frames is now T/τ ($\tau = 1$), where T is the relative cosmic time between them. The transformation obtained is the cosmological transformation given in Section 2.4.

And finally, for unchanged x's we have a two-dimensional rotation $(t, v) \rightarrow (t', v')$, that has not been discussed before. The frame is now fixed at a point in space and the parameter of the transformation is $X/R = \tan\psi$, where ψ is the angle of rotation in the $t - v$ plane, and $R = c\tau (= 1)$.

From the above discussion one can easily find the most general transformation that includes the above three subtransformations. This can be done, for example, like in the three-dimensional successive rotations in classical mechanics. We will not go through that here even though it is not complicated.

2.11.2   Electrodynamics in five dimensions

One can introduce in these five dimensions a skew-symmetric tensor $f_{\mu\nu}$ like in electrodynamics, with μ, ν=0,1,2,3,4. For μ, ν=0,1,2,3, this is exactly the electromagnetic field. The field (H, **W**) describes a new interaction. Accordingly, $f_{\mu\nu}$ is presented in the form

$$
f_{\mu\nu} = \begin{pmatrix}
0 & -E_x & -E_y & -E_z & H \\
E_x & 0 & H_z & -H_y & W_x \\
E_y & -H_z & 0 & H_x & W_y \\
E_z & H_y & -H_x & 0 & W_z \\
-H & -W_x & -W_y & -W_z & 0
\end{pmatrix}. \tag{2.42}
$$

Here **E** and **H** are the electric and magnetic fields.

It is well known that the classification with respect to strengths of fields and particles leads to four types of interactions: strong, electromagnetic, weak and gravitational. This classification, however, has become mixed up since the strengths of the interactions are energy dependent. At high energy the strong interaction becomes weaker and roughly equal to the electromagnetic interactions. There are also interactions of the same strength as the weak interaction, called the Higgs interactions that involve a zero mass with spin 0 particle and a W or Z particle. Moreover, there are interactions that involve a photon and a Higgs particle, with the possibility of a W or Z particle. And so one can talk about the electromagnetic forces due to photons in a very limited sense. So why not unifying the Higgs and the electromagnetic interactions at high energy? This has not been possible up to recently because there was no framework that is suitable to do it. But now it is possible within the framework of cosmology.

Our interpretation for the scalar H and the vector $\mathbf{W}$ is that they are the Higgs massless particle and the W or the Z particle. In this way the Higgs interaction is unified with the electromagnetic interaction within the framework of cosmology, where the Higgs interaction is associated with the expansion of the Universe.

2.11.3 Field equations

What is needed is to write the field equations which are assumed to be given by

$$\frac{\partial f^{\alpha\beta}}{\partial x^{\beta}} = 4\pi j^{\alpha}, \tag{2.43}$$

$$\frac{\partial f_{\alpha\beta}}{\partial x^{\gamma}} + \frac{\partial f_{\beta\gamma}}{\partial x^{\alpha}} + \frac{\partial f_{\gamma\alpha}}{\partial x^{\beta}} = 0, \tag{2.44}$$

where j^{α} is the current. For $\alpha = 0, 1, 2, 3$ it is the ordinary electric current, and for $\alpha = 4$ it is an additional component, with $j^{0} = \rho$ is the charge density, j^{m} (m=1,2,3) =$\mathbf{j}$ is the vector current density, and j^{4} is the velocity-component charge density denoted by $\tilde{\rho}$.

We introduce the potential A_{μ} in five dimensions as in electrodynamics by

$$f_{\mu\nu} = \frac{\partial A_{\mu}}{\partial x^{\nu}} - \frac{\partial A_{\nu}}{\partial x^{\mu}}, \tag{2.45}$$

with $A_{\mu} = (A_0, A_m, A_4) = (\phi, -\mathbf{A}, \tilde{\phi})$, where ϕ and $\mathbf{A}$ are the scalar and vector electromagnetic potentials.

2.12 Generalized Maxwell's equations

A straightforward calculation, using Eq. (2.45), gives

$$\mathbf{E} = -\nabla\phi - \frac{\partial \mathbf{A}}{\partial t}, \tag{2.46a}$$

$$\mathbf{H} = \nabla \times \mathbf{A}, \tag{2.46b}$$

$$\mathbf{W} = -\nabla\tilde{\phi} - \frac{\partial\mathbf{A}}{\partial v}, \tag{2.46c}$$

$$H = \frac{\partial\phi}{\partial v} - \frac{\partial\tilde{\phi}}{\partial t}. \tag{2.46d}$$

From Eqs. (2.43) and (2.44) we get the rest of the generalized Maxwell's equations:

$$\nabla \cdot \mathbf{E} + \frac{\partial H}{\partial v} = 4\pi\rho, \tag{2.47a}$$

$$\nabla \times \mathbf{E} = -\frac{\partial\mathbf{H}}{\partial t}, \tag{2.47b}$$

$$\nabla \cdot \mathbf{H} = 0, \tag{2.47c}$$

$$\nabla \times \mathbf{H} = \frac{\partial\mathbf{E}}{\partial t} + \frac{\partial\mathbf{W}}{\partial v} + 4\pi\mathbf{j}, \tag{2.47d}$$

$$\nabla \cdot \mathbf{W} = \frac{\partial H}{\partial t} + 4\pi\tilde{\rho}, \tag{2.47e}$$

$$\nabla \times \mathbf{W} = \frac{\partial\mathbf{H}}{\partial v}. \tag{2.47f}$$

The wave equation for A_μ is then given by

$$\left(\frac{\partial^2}{\partial t^2} - \nabla^2 + \frac{\partial^2}{\partial v^2}\right) A_\mu = -j_\mu, \tag{2.48}$$

with the condition $\partial A^\alpha/\partial x^\alpha = 0$. There are now a delay in time and in velocity in the solutions of this equation.

2.12.1   The mix-up

It is important to examine the mix-up between the different fields under the transformation in five dimensions. We recall that we have three subtransformations. These are the Lorentz, the cosmological and the 2-rotation. They are given by

$$
L = \frac{1}{\sqrt{1 - v^2/c^2}}
\begin{pmatrix}
1 & -v/c & 0 & 0 & 0 \\
-v/c & 1 & 0 & 0 & 0 \\
0 & 0 & 1 & 0 & 0 \\
0 & 0 & 0 & 1 & 0 \\
0 & 0 & 0 & 0 & 1
\end{pmatrix},
\qquad (2.49)
$$

where $y' = y, \ z' = z, \ v' = v,$

$$
C = \frac{1}{\sqrt{1 - t^2/\tau^2}}
\begin{pmatrix}
1 & 0 & 0 & 0 & 0 \\
0 & 1 & 0 & 0 & -t/\tau \\
0 & 0 & 1 & 0 & 0 \\
0 & 0 & 0 & 1 & 0 \\
0 & -t/\tau & 0 & 0 & 1
\end{pmatrix},
\qquad (2.50)
$$

where $t' = t, \ y' = y, \ z' = z,$ and

$$
R =
\begin{pmatrix}
\cos\psi & 0 & 0 & 0 & \sin\psi \\
0 & 1 & 0 & 0 & 0 \\
0 & 0 & 1 & 0 & 0 \\
0 & 0 & 0 & 1 & 0 \\
-\sin\psi & 0 & 0 & 0 & \cos\psi
\end{pmatrix},
\qquad (2.51)
$$

where $x' = x, \ y' = y, \ z' = z,$ and ψ is the angle between the time and velocity axes.

To find out what are the transformed quantities, it is convenient to use the contravariant components of $f_{\mu\nu}$ which are easily found to be

$$
f^{\mu\nu} =
\begin{pmatrix}
0 & E_x & E_y & E_z & H \\
-E_x & 0 & H_z & -H_y & -W_x \\
-E_y & -H_z & 0 & H_x & -W_y \\
-E_z & H_y & -H_x & 0 & -W_z \\
-H & W_x & W_y & W_z & 0
\end{pmatrix}.
\qquad (2.52)
$$

One finds, for example, that under the Lorentz transformation $\mathbf{E}' = \mathbf{E}$. We also find that under the cosmological transformation, $\mathbf{W}' = \mathbf{W}$. And under the 2-rotation we find

$$\mathbf{E}' = \mathbf{E}\cos\psi + \mathbf{W}\sin\psi, \qquad (2.53\text{a})$$

$$\mathbf{H}' = \mathbf{H}, \qquad (2.53\text{b})$$

$$H' = H \qquad (2.53\text{c})$$

$$\mathbf{W}' = \mathbf{W}\cos\psi - \mathbf{E}\sin\psi. \qquad (2.53\text{d})$$

2.12.2   Does the Cabibbo angle describe a rotation in the time-velocity plane?

Finally, we raise the question whether our angle ψ is related to the Cabibbo angle that also mixes up different states of particles that is known in particle physics but so far has not been understood. Is the Cabibbo angle origin in cosmology?

With this we finished our flat spacetime discussion of cosmology. In the next chapter, gravitation is invoked and the full theory of general relativity is applied to cosmology.

References

M. Carmeli, *Cosmological Special Relativity: The Large-Scale Structure of Space, Time and Velocity*, Second Edition, World Scientific, Singapore, 2002.

M. Carmeli, J. Hartnett and F. Oliveira, "The cosmic time in terms of the redshift," *Foundations of Physics Letters* **19**, 276 - 283 (2006).

M. Veltman, *Facts and Mysteries in Elementary Particle Physics*, World Scientific, Singapore, 2003.

Chapter 3

Elements of General Relativity

In this chapter a brief review of general relativity theory is given. Only the essentials of the theory are given. We begin the discussion with a brief review of Riemannian geometry, followed by a description of the physical foundations of general relativity. These are the principles of equivalence and of general covariance. The gravitational field equations are then derived in a tensorial form. The Schwarzschild metric of Einstein's field equations is derived. Experimental verification of general relativity is subsequently given. The chapter is then concluded with a review of the problem of motion in the gravitational field.

3.1 Riemannian geometry

3.1.1 Transformation of coordinates

Any four independent variables x^μ, where Greek letters take the values 0, 1, 2, 3, may be considered as the coordinates in a four-dimensional space V_4. Each set of values of x^μ defines a point in V_4. Let there be another set of coordinates x'^μ related to the first set x^ν by

$$x'^\mu = f^\mu\left(x^\nu\right),\qquad(3.1)$$

where f^μ are four independent real functions of x^ν. A necessary and sufficient condition that f^μ be independent is that their

Jacobian,

$$\left| \frac{\partial f^\mu}{\partial x^\nu} \right| = \begin{vmatrix} \frac{\partial f^0}{\partial x^0} & \cdots & \frac{\partial f^3}{\partial x^0} \\ \vdots & & \\ \frac{\partial f^0}{\partial x^3} & \cdots & \frac{\partial f^3}{\partial x^3} \end{vmatrix}, \tag{3.2}$$

does not vanish identically. Equation (3.1) defines a *transformation of coordinates* in the space V_4. When the Jacobian is different from zero, one can also write x^μ in terms of x'^ν as

$$x^\mu = g^\mu \left(x'^\nu \right). \tag{3.3}$$

A *direction* at a point P in the space V_4 is determined by the differential dx^μ. The same direction is determined in another set of coordinates x'^μ by the differential dx'^μ. The two differentials are related, using Eq. (3.1), by

$$dx'^\mu = \frac{\partial x'^\mu}{\partial x^\nu} dx^\nu = \frac{\partial f^\mu}{\partial x^\nu} dx^\nu. \tag{3.4}$$

Here the Einstein summation convention is used, according to which repeated Greek indices are summed over the values 0, 1, 2, 3.

3.1.2   Contravariant vectors

Let two sets of functions V^μ and V'^μ be related by

$$V'^\mu = \frac{\partial x'^\mu}{\partial x^\nu} V^\nu, \tag{3.5}$$

similar to the way the differentials dx'^μ and dx^μ are related. V^μ and V'^μ are then called the *components* of a *contravariant vector* in the coordinate systems x^μ and x'^μ, respectively. Hence any four functions of the x's in one coordinate system can be taken as the components of a contravariant vector whose components in any other coordinate system are given by Eq. (3.5).

A contravariant vector determines a direction at each point of the space V_4. Let V^μ be the components of a contravariant vector and let dx^μ be a displacement in the direction of V^μ. Then $dx^0/V^0 = \cdots = dx^3/V^3$. This set of equations admits three independent functions $f^k(x^\mu) = c^k$, where $k = 0, 1, 2$, and the c's are arbitrary constants and the matrix $\partial f^k/\partial c^\mu$ is of rank three. The functions f^k are solutions of the partial differential equation $V^\nu \partial f^k/\partial x^\nu = 0$. Hence, using the transformation laws (3.1) and (3.3), one obtains $V'^k = 0$ for $k = 0, 1, 2$, and $V'^3 \neq 0$. Hence a system of coordinates can be chosen in terms of which all components but one of a given contravariant vector are equal to zero.

3.1.3 Invariants. Covariant vectors

Two functions $f(x)$ and $f'(x')$ define an *invariant* if they are reducible to each other by a coordinate transformation.

Let f be a function of the coordinates. Then

$$\frac{\partial f}{\partial x'^\mu} = \frac{\partial f}{\partial x^\nu}\frac{\partial x^\nu}{\partial x'^\mu}. \tag{3.6}$$

Two sets of functions V_μ and V'_μ are called the components of a *covariant vector* in the systems x and x', respectively, if they are related by a transformation law of the form (3.6),

$$V'_\mu = \frac{\partial x^\nu}{\partial x'^\mu}V_\nu. \tag{3.7}$$

For example, if f is a scalar function, then $\partial f/\partial x^\mu$ is a covariant vector. It is called the *gradient* of f. The product $V^\mu W_\mu$ is an invariant if V is a contravariant vector and W is a covariant vector. Conversely, if the quantity $V^\mu W_\mu$ is an invariant and either V^μ or W_μ are arbitrary vectors, then the other set is a vector.

3.1.4   Tensors

Tensors of any order are defined by generalizing Eqs. (3.5) and (3.7). Thus the equation

$$T'^{\mu_1\cdots\mu_m}_{\nu_1\cdots\nu_n} = \frac{\partial x'^{\mu_1}}{\partial x^{\rho_1}}\cdots\frac{\partial x'^{\mu_m}}{\partial x^{\rho_m}}\frac{\partial x^{\sigma_1}}{\partial x'^{\nu_1}}\cdots\frac{\partial x^{\sigma_n}}{\partial x'^{\nu_n}}T^{\rho_1\cdots\rho_m}_{\sigma_1\cdots\sigma_n} \tag{3.8}$$

defines a mixed tensor of order $m + n$, contravariant of the mth order and covariant of the nth order. If the Kronecker delta function is taken as the components of a mixed tensor of the second order in one set of coordinates, for example, then it defines the components of a tensor in any set of coordinates. An invariant is a tensor of zero order and a vector is a tensor of order one.

When the relative position of two indices, either contravariant or covariant, is immaterial, the tensor is called *symmetric* with respect to these indices. When the relative position of two indices of a tensor is interchanged and the tensor obtained differs only in sign from the original one, the tensor is called *skew-symmetric* with respect to these indices. The process by means of which from a mixed tensor of order r one obtains a tensor of order $r - 2$ is called *contraction*.

3.1.5   Metric tensor

Let $g_{\mu\nu}$ be the components of the *metric tensor*, i.e., a symmetric covariant tensor, which is a function of coordinates, and let $g = \det g_{\mu\nu}$. The quantity $g^{\mu\nu}$, denoting the cofactor of $g_{\mu\nu}$ divided by g, is a symmetric contravariant tensor and satisfies

$$g^{\mu\rho}g_{\nu\rho} = \delta^{\mu}_{\nu}. \tag{3.9}$$

The element of length is defined by means of a quadratic differential form $ds^2 = g_{\mu\nu}dx^{\mu}dx^{\nu}$. By means of the tensors $g_{\mu\nu}$ and $g^{\mu\nu}$ one can lower or raise tensor indices:

$$T^{\mu}{}_{\nu\rho} = g^{\mu\sigma}T_{\sigma\nu\rho}, \tag{3.10a}$$

$$T_\alpha{}^{\beta\gamma} = g_{\alpha\rho}T^{\rho\beta\gamma}. \tag{3.10b}$$

Certain other quantities transform according to the law

$$T'^{\mu\cdots}_{\alpha\cdots} = J^N \frac{\partial x'^\mu}{\partial x^\rho}\frac{\partial x^\beta}{\partial x'^\alpha}\cdots T^{\rho\cdots}_{\beta\cdots}. \tag{3.11}$$

Here J is the Jacobian determinant $|\,\partial x^\alpha/\partial x'^\beta\,|$. The superscript N is the power to which J is raised. $T^{\mu\cdots}_{\nu\cdots}$ is called a *tensor density* of weight N. For example, if g' denotes det $g'_{\mu\nu}$ then $g' = J^2 g$, where $g = $ det $g_{\mu\nu}$. Hence one has for the four-dimensional elements in two coordinate systems the equality:

$$\sqrt{-g}d^4x = \sqrt{-g'}d^4x'. \tag{3.12}$$

3.1.6 Christoffel symbols

From the two tensors $g_{\mu\nu}$ and $g^{\mu\nu}$ one can define the two functions

$$\Gamma_{\alpha\rho\sigma} = \frac{1}{2}\left(\frac{\partial g_{\rho\alpha}}{\partial x^\sigma} + \frac{\partial g_{\sigma\alpha}}{\partial x^\rho} - \frac{\partial g_{\rho\sigma}}{\partial x^\alpha}\right), \tag{3.13}$$

$$\Gamma^\mu_{\rho\sigma} = g^{\mu\alpha}\Gamma_{\alpha\rho\sigma}. \tag{3.14}$$

They are symmetric in ρ and σ, and are called the *Christoffel symbols* of the *first* and *second* kind, respectively.

Both kinds of Christoffel symbols are not components of tensors. By starting with the differential transformation law for $g_{\mu\nu}$ it is not too difficult to show that $\Gamma_{\alpha\rho\sigma}$ transforms according to the following relation:

$$\Gamma'_{\nu\mu\alpha} = \frac{\partial x^\beta}{\partial x'^\mu}\frac{\partial x^\gamma}{\partial x'^\alpha}\frac{\partial x^\delta}{\partial x'^\nu}\Gamma_{\delta\beta\gamma} + g_{\beta\gamma}\frac{\partial x^\beta}{\partial x'^\nu}\frac{\partial^2 x^\gamma}{\partial x'^\mu\partial x'^\alpha}. \tag{3.15}$$

Making use of the transformation law for $g^{\alpha\beta}$ then leads to the transformation law of $\Gamma^\delta_{\beta\nu}$ as

$$\Gamma'^\delta_{\beta\nu} = \frac{\partial x'^\delta}{\partial x^\alpha}\frac{\partial x^\mu}{\partial x'^\beta}\frac{\partial x^\sigma}{\partial x'^\nu}\Gamma^\alpha_{\mu\sigma} + \frac{\partial x'^\delta}{\partial x^\sigma}\frac{\partial^2 x^\sigma}{\partial x'^\beta x'^\nu}. \tag{3.16}$$

From Eq. (3.13) we obtain

$$\Gamma^{\mu}_{\alpha\mu} = \frac{1}{2} g^{\mu\nu} \frac{\partial g_{\mu\nu}}{\partial x^{\alpha}}. \tag{3.17}$$

This equation can be rewritten in terms of the determinant g of $g_{\mu\nu}$. The rule for expansion of a determinant leads to the formula

$$\frac{\partial g}{\partial g_{\mu\nu}} = \Delta^{\mu\nu}, \tag{3.18}$$

where $\Delta^{\mu\nu}$ is the cofactor of the element $g_{\mu\nu}$. From the law for obtaining the inverse of a determinant, and from the definition of $g^{\mu\nu}$, Eq. (3.18) may be written as

$$\frac{\partial g}{\partial g_{\mu\nu}} = g g^{\mu\nu}, \tag{3.19}$$

and consequently

$$dg = g g^{\mu\nu} dg_{\mu\nu} = -g g_{\mu\nu} dg^{\mu\nu}. \tag{3.20}$$

Hence we have

$$\frac{\partial g}{\partial x^{\alpha}} = g g^{\mu\nu} \frac{\partial g_{\mu\nu}}{\partial x^{\alpha}} = -g g_{\mu\nu} \frac{\partial g^{\mu\nu}}{\partial x^{\alpha}}. \tag{3.21}$$

The use of Eq. (3.21) enables us to write Eq. (3.17) in the form

$$\Gamma^{\mu}_{\alpha\mu} = \frac{1}{2g} \frac{\partial g}{\partial x^{\alpha}} = \frac{1}{\sqrt{-g}} \frac{\partial \sqrt{-g}}{\partial x^{\alpha}}. \tag{3.22}$$

3.1.7   Covariant differentiation

We have seen that the derivatives of an invariant are the components of a covariant vector. This is the only case for a general system of coordinates in which the derivative of a tensor is a tensor. However, there are expressions involving first derivatives which are components of a tensor. To see this we proceed as follows.

Let V^μ and V'^ν be a contravariant vector in two coordinate systems x and x'. Then

$$V^\mu = V'^\nu \frac{\partial x^\mu}{\partial x'^\nu}. \tag{3.23}$$

Differentiating this equation with respect to x^α and using Eq. (3.16) gives:

$$\frac{\partial V^\mu}{\partial x^\alpha} = \left(\frac{\partial V'^\rho}{\partial x'^\nu} + V'^\sigma \Gamma'^\rho_{\sigma\nu} \right) \frac{\partial x'^\nu}{\partial x^\alpha} \frac{\partial x^\mu}{\partial x'^\rho} - V^\rho \Gamma^\mu_{\rho\alpha}. \tag{3.24}$$

Hence if we define a *covariant derivative* of V^μ by

$$\nabla_\alpha V^\mu = \partial_\alpha V^\mu + \Gamma^\mu_{\rho\alpha} V^\rho, \tag{3.25}$$

Eq. (3.24) can be written as

$$\nabla_\alpha V^\mu = \nabla_\nu V'^\rho \frac{\partial x'^\nu}{\partial x^\alpha} \frac{\partial x^\mu}{\partial x'^\rho}. \tag{3.26}$$

Therefore $\nabla_\alpha V^\mu$ is a mixed tensor of second order.

In the same way one shows that the covariant derivative of a covariant vector V_μ is given by:

$$\nabla_\alpha V_\mu = \partial_\alpha V_\mu - \Gamma^\rho_{\mu\alpha} V_\rho, \tag{3.27}$$

From the above equation one has for the *curl* of a vector V_μ:

$$\nabla_\beta V_\alpha - \nabla_\alpha V_\beta = \partial_\beta V_\alpha - \partial_\alpha V_\beta. \tag{3.28}$$

Hence a necessary and sufficient condition that the first covariant derivative of a covariant vector be symmetric is that the vector be a gradient.

It is easily seen, using the law of covariant differentiation of tensors, that

$$\nabla_\rho g^{\mu\nu} = 0, \tag{3.29a}$$

$$\nabla_\rho g_{\mu\nu} = 0, \tag{3.29b}$$

$$\nabla_\rho \delta^\mu_\nu = 0. \tag{3.29c}$$

Other properties of covariant differentiation can be established.

3.1.8 Riemann, Ricci and Einstein tensors

If we differentiate covariantly the tensor $\nabla_\alpha V_\mu$, given by Eq. (3.27), we obtain

$$\left(\nabla_\gamma \nabla_\beta - \nabla_\beta \nabla_\gamma\right) V_\alpha = R^\delta{}_{\alpha\beta\gamma} V_\delta, \qquad (3.30)$$

where $R^\delta{}_{\alpha\beta\gamma}$ is called the *Riemann tensor* and is given by

$$R^\delta{}_{\alpha\beta\gamma} = \partial_\beta \Gamma^\delta_{\alpha\gamma} - \partial_\gamma \Gamma^\delta_{\alpha\beta} + \Gamma^\mu_{\alpha\gamma} \Gamma^\delta_{\mu\beta} - \Gamma^\mu_{\alpha\beta} \Gamma^\delta_{\mu\gamma}. \qquad (3.31)$$

A generalization of Eq. (3.30) to an arbitrary tensor can be made.

One can show that in order that there can exist a coordinate system in which the first covariant derivatives reduce to ordinary ones at every point in space, it is necessary and sufficient that the Riemann tensor be zero and that the coordinates be those in which the metric is constant.

One notices that the Riemann tensor satisfies

$$R_{\alpha\beta\gamma\delta} = -R_{\beta\alpha\gamma\delta} = -R_{\alpha\beta\delta\gamma} = R_{\gamma\delta\alpha\beta}, \qquad (3.32a)$$

$$R_{\alpha\beta\gamma\delta} + R_{\alpha\gamma\delta\beta} + R_{\alpha\delta\beta\gamma} = 0. \qquad (3.32b)$$

Moreover, counting the number of components, one finds that in a four-dimensional space the Riemann tensor has 20 components.

From the Riemann tensor one can define the *Ricci tensor* and the *Ricci scalar* by

$$R_{\mu\nu} = R^\alpha{}_{\mu\alpha\nu} = \frac{1}{\sqrt{-g}} \left(\sqrt{-g}\,\Gamma^\alpha_{\mu\nu}\right)_{,\alpha} - \left(\ln\sqrt{-g}\right)_{,\mu\nu} - \Gamma^\alpha_{\mu\beta}\Gamma^\beta_{\nu\alpha},$$
$$(3.33)$$
$$R = R^\mu{}_\mu, \qquad (3.34)$$

respectively. Here a comma denotes partial differentiation, $f_{,\alpha} = \partial f/\partial x^\alpha$. The *Einstein tensor* is then defined by

$$G_{\mu\nu} = R_{\mu\nu} - \frac{1}{2} g_{\mu\nu} R. \qquad (3.35)$$

The last important tensor constructed from the Riemann tensor is the *Weyl conformal tensor*:

$$C_{\rho\sigma\mu\nu} = R_{\rho\sigma\mu\nu} - \frac{1}{2}\left(g_{\rho\mu}R_{\nu\sigma} - g_{\rho\nu}R_{\mu\sigma} - g_{\sigma\mu}R_{\nu\rho} + g_{\sigma\nu}R_{\mu\rho}\right)$$

$$-\frac{1}{6}\left(g_{\rho\nu}g_{\mu\sigma} - g_{\rho\mu}g_{\nu\sigma}\right)R. \tag{3.36}$$

It has the special property that its trace is zero,

$$C^{\rho}{}_{\mu\rho\nu} = 0. \tag{3.37}$$

Furthermore, if the Weyl tensor vanishes everywhere, then the metric is said to be *conformally flat*. (Two spaces V and $\tilde{V}$ are called conformal spaces if their metric tensors $g_{\mu\nu}$ and $\tilde{g}_{\mu\nu}$ are related by $\tilde{g}_{\mu\nu} = e^{\beta}g_{\mu\nu}$, where β is a function of the coordinates.) That is, there exists a mapping such that $g_{\mu\nu}$ can be diagonalized, with $\pm\beta(x)$ appearing in the diagonal positions, and where $\beta(x)$ is some function. This follows from the fact that the Weyl tensor can be expressed entirely in terms of the density $\tilde{g}_{\mu\nu} = (-g)^{-1/4}g_{\mu\nu}$ and its inverse, and is equal to the Riemann tensor formed by replacing $g_{\mu\nu}$ by $\tilde{g}_{\mu\nu}$, $R_{\alpha\beta\gamma\delta}(\tilde{g}_{\mu\nu}) = C_{\alpha\beta\gamma\delta}(g_{\mu\nu})$.

Consequently, the vanishing of the Weyl tensor implies the vanishing of $R_{\alpha\beta\gamma\delta}(\tilde{g}_{\mu\nu})$, which in turn implies that there exists a mapping such that $\tilde{g}_{\mu\nu}$ is everywhere diagonal, with ± 1 appearing along the diagonal. Only g is arbitrary and $\pm(-g)^{1/4}$ appears along the diagonal of $g_{\mu\nu}$.

3.1.9 Geodesics

The differential equations of the curves of extremal length are called *geodesic equations*. To find their equations we seek the relations which must be satisfied to give a stationary value to the integral $\int ds$. Hence we have to find the solution of the variational problem

$$\delta \int L \, ds = 0, \tag{3.38}$$

where the Lagrangian L is given by

$$L = \left(g_{\mu\nu} \frac{dx^\mu}{ds} \frac{dx^\nu}{ds} \right)^{1/2}. \tag{3.39}$$

Accordingly we have

$$\delta \int L \, ds = \int \left[\frac{\partial L}{\partial x^\mu} \delta x^\mu + \frac{\partial L}{\partial (dx^\mu/ds)} \delta \left(\frac{dx^\mu}{ds} \right) \right] ds. \tag{3.40}$$

The second term of the integrand may be written as the two terms

$$\frac{d}{ds} \left[\frac{\partial L}{\partial (dx^\mu/ds)} \delta x^\mu \right] - \frac{d}{ds} \left[\frac{\partial L}{\partial (dx^\mu/ds)} \right] \delta x^\mu. \tag{3.41}$$

On integration, the first of these expressions contributes nothing since the variations are assumed to vanish at the end points of the curve.

As expected, the equation obtained is the usual Lagrange equation:

$$\frac{d}{ds} \frac{\partial L}{\partial (dx^\mu/ds)} - \frac{\partial L}{\partial x^\mu} = 0. \tag{3.42}$$

A simple calculation then gives, using the Lagrangian given by Eq. (3.39),

$$\frac{d^2 x^\mu}{ds^2} + \Gamma^\mu_{\alpha\beta} \frac{dx^\alpha}{ds} \frac{dx^\beta}{ds} = 0. \tag{3.43}$$

3.1.10 Bianchi identities

A study of Eq. (3.16) shows that it is always possible to choose a coordinate system in which all the Christoffel symbols vanish at a point. For, suppose the Christoffel symbols do not vanish at a point A. One can then carry out the coordinate transformation

$$x'^\alpha = x^\alpha - x^\alpha_A + \frac{1}{2} \Gamma^\alpha_{\beta\gamma} (A) \left(x^\beta - x^\beta_A \right) \left(x^\gamma - x^\gamma_A \right), \tag{3.44}$$

where the subscript A indicates to the value at the point A. By Eq. (3.16) one finds that the Christoffel symbols in the new coordinate system vanish at the point A.

A coordinate system for which the Christoffel symbols vanish at a point is called *geodesic*. (It is also possible to transform away the Christoffel symbols along a given curve.)

If we choose a geodesic coordinate system at a point A, then at A one has

$$\nabla_\nu R^\mu{}_{\delta\beta\gamma} = \partial_\beta \partial_\nu \Gamma^\mu_{\delta\gamma} - \partial_\gamma \partial_\nu \Gamma^\mu_{\delta\beta}. \tag{3.45}$$

Consequently, at the point A one has:

$$\nabla_\nu R^\mu{}_{\delta\beta\gamma} + \nabla_\gamma R^\mu{}_{\delta\nu\beta} + \nabla_\beta R^\mu{}_{\delta\gamma\nu} = 0. \tag{3.46}$$

Since the terms of this equation are components of a tensor, this equation holds for any coordinate system and at each point. Hence Eq. (3.46) is an identity throughout the space. It is known as the *Bianchi identities*.

Multiplication of Eq. (3.46) by $g^{\delta\beta}\delta^\gamma_\mu$ gives

$$g^{\delta\beta}\left(\nabla_\nu R^\gamma{}_{\delta\beta\gamma} + \nabla_\gamma R^\gamma{}_{\delta\nu\beta} + \nabla_\beta R^\gamma{}_{\delta\gamma\nu}\right) = 0. \tag{3.47}$$

Using the symmetry properties of the Riemann tensor, the last equation becomes:

$$\nabla_\nu \left(R_\gamma{}^\nu - \frac{1}{2}\delta^\nu_\gamma R\right) = \nabla_\nu G_\gamma{}^\nu = 0. \tag{3.48}$$

Equation (3.48) is called the *contracted Bianchi identity*.

After having developed the mathematical tools to describe general relativity theory, we now turn to the physical foundations of the theory.

3.2 Principle of equivalence

3.2.1 Null experiments. Eötvös experiment

One of the most interesting *null experiments* in physics is due to Eötvös, first performed in 1890 and recently repeated by Dicke.

The experiment showed, in great precision, that all bodies fall with the same acceleration. The roots of the experiment go back to Newton and Galileo, who demonstrated experimentally that the gravitational acceleration of a body is independent of its composition.

The importance of the Eötvös experiment is in the fact that the null result of the experiment is a *necessary* condition for the theory of general relativity to be valid.

Eötvös employed a static torsion balance, balancing a component of the Earth's gravitational pull on the weight against the centrifugal force field of the Earth acting on the weight. He employed a horizontal torsion beam, 40 cm long, suspended by a fine wire. From the ends of the torsion beam were suspended two masses of different compositions, one lower than the other. A lack of exact proportionality between the inertial and gravitational masses of the two bodies would then lead to a torque tending to rotate the balance. There appears to be no need for the one mass to be suspended lower than the other.

The experiment of Eötvös showed, with an accuracy of a few parts in 10^9, that inertial and gravitational masses are equal.

In the experiment performed by Dicke, the gravitational acceleration toward the Sun of small gold and aluminum weights were compared and found to be equal with an accuracy of about one part in 10^{11}. Hence the necessary condition to be satisfied for the validity of general relativity theory seems to be rather satisfactorily met.

The question therefore arises as to what extent is this experiment also a sufficient condition to be satisfied in order that general relativity theory be valid.

It has been emphasized by Dicke that gold and aluminum differ from each other rather greatly in several important aspects. First, the neutron to proton ratio is quite different in the two elements, varying from 1.08 in aluminum to 1.50 in gold. Second, the electrons in aluminum move with nonrelativistic velocities, but in gold the k-shell electrons have a 15 per cent increase in their masses as a result of their relativistic velocities. Third, the electromagnetic negative contribution to the binding energy of the nucleus varies as z^2 and represents $\frac{1}{2}$ per cent of the total mass of a gold atom, whereas it is negligible in aluminum. Fourth, the virtual pair field and other fields would be expected to be different in the two atoms. We thus conclude that the physical aspects of gold and aluminum differ substantially, and consequently the equality of their accelerations represents an important condition to be satisfied by any theory of gravitation.

Since the accuracy of the Eötvös experiment is great, the question arises as to whether it implies that the equivalence principle is very nearly valid. This is true in a limited sense; certain aspects of the equivalence principle are not supported in the slightest by the Eötvös experiment.

3.3   Principle of general covariance

We have seen in the preceding section that a gravitational field can be considered locally equivalent to an accelerated frame. This implies that the special theory of relativity cannot be valid in an extended region where gravitational fields are present. A curved spacetime is needed and all laws of nature should be covariant under the most general coordinate transformations.

The original formulation of general relativity by Einstein was based on two principles: (1) the principle of equivalence (discussed in detail in the last section); and (2) the principle of general covariance.

The principle of general covariance is often stated in one of the following forms, which are not exactly equivalent:

(1) All coordinate systems are equally good for stating the laws of physics, and they should be treated on the same footing.

(2) The equations of physics should have tensorial forms.

(3) The equations of physics should have the same form in all coordinate systems.

According to the principle of general covariance, the coordinates become nothing more than a bookkeeping system to label the events. The principle is a valuable guide to deducing correct equations.

It has been pointed out that any spacetime physical law can be written in a covariant form and hence the principle of general covariance has no necessary physical consequences, and Einstein concurred with this view.

In spite of Einstein's acceptance of this objection, it appears that the principle of general covariance was introduced by Einstein as a generalization of the principle of special relativity and he often referred to it as the principle of general relativity. In fact the principle of equivalence (which necessarily leads to the introduction of a curved spacetime), plus the assumption of general covariance, is most of what is needed to generate Einstein's theory of general relativity. They lead directly to the idea that gravitation can be explained by means of Riemannian geometry. This is done in the next section.

3.4   Gravitational field equations

We have seen in Section 3.1 that the Riemannian geometry is characterized by a geometrical metric, i.e., a symmetric tensor $g_{\mu\nu}$ from which one can construct other quantities. Classical general relativity theory identifies this tensor as the gravitational potential. Hence in general relativity there are ten components to the gravitational potential, as compared with the single potential function in the Newtonian theory of gravitation.

3.4.1 Einstein's field equations

In trying to arrive at the desired gravitational field equations that the metric tensor has to satisfy, we are guided by the requirement that, in an appropriate limit, the theory should be reduced to the Newtonian gravitational theory. In the latter theory, the gravitational potential ϕ is determined by the Poisson equation:

$$\nabla^2 \phi = 4\pi G \rho, \tag{3.49}$$

where G $(= 6.67 \times 10^{-8} \text{ cm}^3 \text{ gm}^{-1} \text{ sec}^{-2})$ is the Newton gravitational constant and ρ is the mass density of matter. Hence $g_{\mu\nu}$ should satisfy second order partial differential equations. The equations should then be related to the energy-momentum tensor $T_{\mu\nu}$ linearly. Such equations are

$$R_{\mu\nu} - \frac{1}{2} g_{\mu\nu} R = \kappa T_{\mu\nu}, \tag{3.50}$$

where κ is some constant to be determined. In cosmology theory, one sometimes adds an additional term, $\lambda g_{\mu\nu}$, to the left-hand side of Eq. (3.50). The constant λ is known as a cosmological constant.

But the contracted Bianchi identities, Eq. (3.48), show that the covariant divergence of the left-hand side of Eq. (3.50) vanishes. Hence

$$\nabla_\nu T_\mu{}^\nu = 0, \tag{3.51}$$

which expresses the covariant conservation of energy and momentum. The constant κ can be determined by going to the limit of weak gravitational field. Its value is $\kappa = 8\pi G/c^4$. The constant κ is known as Einstein's gravitational constant.

3.4.2 Deduction of Einstein's equations from variational principle

We start with the action integral

$$I = \int \sqrt{-g}\,(L_G - 2\kappa L_F)\,d^4x, \tag{3.52}$$

and demand its variation to be zero. Here L_G and L_F are the Lagrangians for the gravitational and other fields, respectively. We take $L_G = R$, where R is the Ricci scalar, $R = R_{\mu\nu}g^{\mu\nu}$.

The first part of the integral (3.52) gives

$$\delta \int \sqrt{-g}\,R\,d^4x = \int \sqrt{-g}\,g^{\mu\nu}\delta R_{\mu\nu}\,d^4x + \int R_{\mu\nu}\delta\left(\sqrt{-g}\,g^{\mu\nu}\right)d^4x. \tag{3.53}$$

To find $\delta R_{\mu\nu}$ we note that in a geodesic coordinate system one has

$$\delta R_{\mu\nu} = \nabla_\alpha\left(\delta\Gamma^\alpha_{\mu\nu}\right) - \nabla_\nu\left(\delta\Gamma^\alpha_{\mu\alpha}\right). \tag{3.54}$$

But the latter is a tensorial equation. Hence it is valid in all coordinate systems. Consequently, the first integral on the right-hand side of Eq. (3.53) can be written as

$$\int \sqrt{-g}\,g^{\mu\nu}\delta R_{\mu\nu}\,d^4x = \int \sqrt{-g}\,\nabla_\alpha\left(g^{\mu\nu}\delta\Gamma^\alpha_{\mu\nu} - g^{\mu\alpha}\delta\Gamma^\beta_{\mu\beta}\right)d^4x, \tag{3.55}$$

and hence is equal to

$$\int \frac{\partial}{\partial x^\alpha}\left[\sqrt{-g}\left(g^{\mu\nu}\delta\Gamma^\alpha_{\mu\nu} - g^{\mu\alpha}\delta\Gamma^\beta_{\mu\beta}\right)\right]d^4x. \tag{3.56}$$

This integral, however, vanishes since by Gauss' theorem it is equal to a surface integral which is equal to zero in consequence of the vanishing of the variations at the boundary.

The second integral on the right-hand side of Eq. (3.53) gives, by Eq. (3.21),

$$\int R_{\mu\nu}\delta\left(\sqrt{-g}\,g^{\mu\nu}\right)d^4x = \int \sqrt{-g}\left(R_{\mu\nu} - \frac{1}{2}g_{\mu\nu}R\right)\delta g^{\mu\nu}\,d^4x. \tag{3.57}$$

The second part of the integral (3.52) leads to

$$\delta \int \sqrt{-g} L_F d^4 x = -\frac{1}{2} \int \sqrt{-g} T_{\mu\nu} \delta g^{\mu\nu} d^4 x, \qquad (3.58)$$

where $T_{\mu\nu}$ is the energy-momentum tensor and is given by

$$T_{\mu\nu} = \frac{-2}{\sqrt{-g}} \left[\left(\frac{\partial \left(\sqrt{-g} L_F \right)}{\partial g^{\mu\nu}_{,\alpha}} \right)_{,\alpha} - \frac{\partial \left(\sqrt{-g} L_F \right)}{\partial g^{\mu\nu}} \right], \qquad (3.59)$$

and a comma denotes partial differentiation, $f_{,\alpha} = \partial_\alpha f$. Combining Eqs. (3.52), (3.57) and (3.58) then leads to the field equations (3.50):

$$R_{\mu\nu} - \frac{1}{2} g_{\mu\nu} R = \kappa T_{\mu\nu}. \qquad (3.60)$$

3.4.3 The electromagnetic energy-momentum tensor

The energy-momentum tensor $T_{\mu\nu}$ for the electromagnetic field is obtained from the general expression (3.59) with the field Lagrangian L_F given by

$$L_F = -\frac{1}{16\pi} g^{\alpha\mu} g^{\beta\nu} f_{\alpha\beta} f_{\mu\nu}. \qquad (3.61)$$

It can easily be shown to be given by

$$T_{\rho\sigma} = \frac{1}{4\pi} \left(\frac{1}{4} g_{\rho\sigma} f_{\alpha\beta} f^{\alpha\beta} - f_{\rho\alpha} f_\sigma{}^\alpha \right). \qquad (3.62)$$

If we calculate the trace of the energy-momentum tensor (3.62) we find that it vanishes,

$$T = T_\rho{}^\rho = g^{\rho\sigma} T_{\rho\sigma} = 0. \qquad (3.63)$$

Using now $R = -\kappa T$ then leads to the vanishing of the Ricci scalar curvature, $R = 0$. We therefore obtain

$$R_{\mu\nu} = \kappa T_{\mu\nu} \qquad (3.64)$$

for the Einstein field equations in the presence of an electromagnetic field. In Eq. (3.64) the energy-momentum tensor $T_{\mu\nu}$ is given by Eq. (3.62).

The Einstein field equations (3.64) and the Maxwell equations constitute *the coupled Einstein-Maxwell field equations.*

3.5   The Schwarzschild metric

In spite of the nonlinearity of the Einstein field equations, there are numerous exact solutions to these equations. Moreover, there are other solutions which are not exact but approximate. Exact solutions are usually obtained using special methods.

The simplest, but of most importance in astrophysics, of all exact solutions to Einstein's field equations is that of Schwartzschild. The solution is *spherically symmetric* and *static.* Such a field can be produced by a spherically symmetric distribution and motion of matter. It follows that the requirement of spherical symmetry alone is sufficient to yield a static solution.

The spherical symmetry of the metric means that the expression for the interval $ds = (g_{\mu\nu}dx^\mu dx^\nu)^{1/2}$ must be the same for all points located at the same distance from the center. In flat space their distance is equal to the radius vector, and the metric is given by (c is taken as equal to 1):

$$ds^2 = dt^2 - dr^2 - r^2 \left(d\theta^2 + \sin^2\theta d\phi^2\right). \qquad (3.65)$$

In a non-Euclidean space, such as the Riemannian one we have in the presence of a gravitational field, there is no quantity which has all the properties of the flat space radius vector, such as that it is equal both to the distance from the center and to the length of the circumference divided by 2π. Therefore, the choice of a radius vector is arbitrary here.

When a mass with spherical symmetry is introduced at the origin, the flat space line element (3.65) must be modified but in a way that retains spherical symmetry. The most general spherically symmetric expression for ds^2 is

$$ds^2 = a\left(r,t\right) dt^2 + b\left(r,t\right) dr^2 + c\left(r,t\right) drdt$$

$$+ d\left(r,t\right) \left(d\theta^2 + \sin^2\theta d\phi^2\right). \tag{3.66}$$

Because of the arbitrariness in the choice of the coordinate system in general relativity theory, we can perform a coordinate transformation which does not destroy the spherical symmetry of ds^2. Hence we can choose new coordinates r' and t' expressed by some functions $r' = r'\left(r,t\right)$ and $t' = t'\left(r,t\right)$.

Making use of these transformations, we can choose the new coordinates so that the coefficient $c\left(r,t\right)$ of the mixed term $drdt$ vanishes and the coefficient $d\left(r,t\right)$ of the angular part to be $-r'^2$, in the metric (3.66). The latter condition implies that the radius vector is now defined in such a way that the circumference of a circle whose center is at the origin of the coordinates is equal to $2\pi r$. It is convenient to express the functions $a\left(r,t\right)$ and $b\left(r,t\right)$ in exponential forms, e^ν and $-e^\lambda$, respectively, where ν and λ are functions of the new coordinates r' and t'. Consequently, the line element (3.66) will have the form

$$ds^2 = e^\nu dt^2 - e^\lambda dr^2 - r^2 \left(d\theta^2 + \sin^2\theta d\phi^2\right), \tag{3.67}$$

where, for brevity, we have dropped the primes from the new coordinates r' and t', and the speed of light c is taken as equal to 1.

We now denote the coordinates t, r, θ, ϕ by x^0, x^1, x^2, x^3, respectively. Hence the components of the covariant metric tensor are given by:

$$g_{\mu\nu} = \begin{pmatrix} e^\nu & 0 & 0 & 0 \\ 0 & -e^\lambda & 0 & 0 \\ 0 & 0 & -r^2 & 0 \\ 0 & 0 & 0 & -r^2\sin^2\theta \end{pmatrix}, \tag{3.68a}$$

whereas those of the contravariant metric tensor are:

$$
g^{\mu\nu} = \begin{pmatrix} e^{-\nu} & 0 & 0 & 0 \\ 0 & -e^{-\lambda} & 0 & 0 \\ 0 & 0 & -r^{-2} & 0 \\ 0 & 0 & 0 & -r^{-2}\sin^{-2}\theta \end{pmatrix}. \tag{3.68b}
$$

To find out the differential equations that the functions ν and λ have to satisfy, according to Einstein's field equations, we first need to calculate the Christoffel symbols associated with the metric (3.68). The nonvanishing components are:

$$
\Gamma^0_{00} = \frac{\dot{\nu}}{2}, \qquad \Gamma^0_{10} = \frac{\nu'}{2}, \qquad \Gamma^0_{11} = \frac{\dot{\lambda}}{2}e^{\lambda-\nu}, \tag{3.69a}
$$

$$
\Gamma^1_{00} = \frac{\nu'}{2}e^{\nu-\lambda}, \qquad \Gamma^1_{10} = \frac{\dot{\lambda}}{2}, \qquad \Gamma^1_{11} = \frac{\lambda'}{2}, \tag{3.69b}
$$

$$
\Gamma^1_{22} = -re^{-\lambda}, \qquad \Gamma^1_{33} = -r\sin^2\theta\, e^{-\lambda}, \qquad \Gamma^2_{12} = \frac{1}{r}, \tag{3.69c}
$$

$$
\Gamma^2_{33} = -\sin\theta\cos\theta, \qquad \Gamma^3_{13} = \frac{1}{r}, \qquad \Gamma^3_{23} = \cot\theta, \tag{3.69d}
$$

where dots and primes denote differentiation with respect to t and r, respectively.

With these Christoffel symbols, we compute the following expressions for the nonvanishing components of the Einstein tensor:

$$
G_0{}^0 = -e^{-\lambda}\left(\frac{1}{r^2} - \frac{\lambda'}{r}\right) + \frac{1}{r^2} = \kappa T_0{}^0, \tag{3.70a}
$$

$$
G_0{}^1 = -e^{-\lambda}\frac{\dot{\lambda}}{r} = \kappa T_0{}^1, \tag{3.70b}
$$

$$
G_1{}^1 = -e^{-\lambda}\left(\frac{\nu'}{r} + \frac{1}{r^2}\right) + \frac{1}{r^2} = \kappa T_1{}^1, \tag{3.70c}
$$

$$G_2{}^2 = -\frac{1}{2}e^{-\lambda}\left(\nu'' + \frac{\nu'^2}{2} + \frac{\nu' - \lambda'}{r} - \frac{\nu'\lambda'}{2}\right)$$

$$+\frac{1}{2}e^{-\nu}\left(\ddot{\lambda} + \frac{\dot{\lambda}^2}{2} - \frac{\dot{\lambda}\dot{\nu}}{2}\right) = \kappa T_2{}^2, \tag{3.70d}$$

$$G_3{}^3 = G_2{}^2 = \kappa T_3{}^3. \tag{3.70e}$$

All other components vanish identically.

The gravitational field equations can now be integrated exactly for the spherical symmetric field in vacuum, i.e., outside the masses producing the field. Setting Eqs. (3.70) equal to zero leads to the independent equations:

$$e^{-\lambda}\left(\frac{\nu'}{r} + \frac{1}{r^2}\right) - \frac{1}{r^2} = 0, \tag{3.71a}$$

$$e^{-\lambda}\left(\frac{\lambda'}{r} - \frac{1}{r^2}\right) + \frac{1}{r^2} = 0, \tag{3.71b}$$

$$\dot{\lambda} = 0. \tag{3.71c}$$

From Eq. (3.71a) and (3.71b) we find $\nu' + \lambda' = 0$, so that $\nu + \lambda = f(t)$, where $f(t)$ is a function of t alone. If we perform now the coordinate transformation $x^0 = h(x'^0)$, $x^k = x'^k$, then $g'_{00} = \dot{h}^2 g_{00}$. Such a transformation amounts to adding to the function ν an arbitrary function of time, while leaving unaffected the other components of the metric. Hence we can choose the function h so that $\nu + \lambda = 0$. Consequently, we see, by Eq. (3.71c), that both ν and λ are time-independent. In other words the spherically symmetric gravitational field in vacuum is automatically static.

Equation (3.71b) can now be integrated. It gives:

$$e^{-\lambda} = e^{\nu} = 1 - \frac{K}{r}, \tag{3.72}$$

where K is an integration constant. We see that for $r \to \infty$, $e^{-\lambda} = e^{\nu} = 1$, i.e., far from the gravitational bodies, the metric reduces to that of the flat space (3.65). The constant K can easily be determined from the requirement that Newton's law of motion be obtained at large distances from the central mass. From the geodesic equation it follows that the radial acceleration of a small test mass at rest with respect to the central mass is:

$$-\Gamma^1_{00} = -\frac{1}{2}\left(1 - \frac{K}{r}\right)\frac{K}{r^2} \rightarrow -\frac{K}{2r^2}. \tag{3.73}$$

Comparing this expression with the Newtonian value $-Gm/r^2$ gives $K = 2Gm$, where m is the central mass and G is the Newton constant.

The constant $2Gm$, or $2Gm/c^2$ in units where c is not taken as equal to 1, is often called the *Schwarzschild radius* of the mass m. For example, the Schwarzschild radius for the Sun is 2.95 km, that for the Earth is 8.9 mm and for an electron is 13.5×10^{-56} cm.

We therefore obtain for the spherically symmetric metric the form:

$$g_{\mu\nu} = \begin{pmatrix} 1 - 2Gm/r & 0 & 0 & 0 \\ 0 & -(1 - 2Gm/r)^{-1} & 0 & 0 \\ 0 & 0 & -r^2 & 0 \\ 0 & 0 & 0 & -r^2\sin^2\theta \end{pmatrix}. \tag{3.74}$$

It is known as the *Schwarzschild metric* and describes the most general spherically symmetric solution of the Einstein field equations in a region of space where the energy-momentum tensor $T^{\mu\nu}$ vanishes. Although $g_{\mu\nu}$ goes to the flat space metric when r goes to infinity, it was *not* necessary to require this asymptotic behavior to obtain the solution.

It is worth mentioning that all spherically symmetric solutions of the Einstein field equations in vacuum which satisfy the boundary conditions at infinity mentioned above are equivalent to the Schwarzschild field, i.e., their time-dependence can be eliminated by a suitable coordinate transformation. This result is due to Birkhoff.

Finally, it is convenient to introduce Cartesian coordinates by means of the coordinate transformation

$$\begin{aligned}
x^1 &= r \sin\theta \cos\phi, \\
x^2 &= r \sin\theta \sin\phi, \\
x^3 &= r \cos\theta.
\end{aligned} \tag{3.75}$$

In terms of these coordinates, the Schwarzschild metric (3.74) will then have the form

$$g_{00} = 1 - \frac{2Gm}{r},$$

$$g_{0r} = 0, \tag{3.76}$$

$$g_{rs} = -\delta_{rs} - \frac{2Gm/r}{1 - 2Gm/r} \frac{x^r x^s}{r^2}.$$

3.6 Experimental tests of general relativity

Up to a few years ago, general relativity was verified by three tests: the gravitational red shift, the deflection of light near massive bodies and the planetary orbit effect on the planets. The first could also be explained, in fact, without the use of the Einstein field equations. However, this picture has been changed.

3.6.1 Gravitational red shift

Consider the clocks at rest at two points 1 and 2. The rate of change of times at these points are then given by $ds\,(1) = \sqrt{g_{00}\,(1)}dt$ and $ds\,(2) = \sqrt{g_{00}\,(2)}dt$. The relation between the rates of identical clocks in a gravitational field is therefore given by $\sqrt{g_{00}(2)/g_{00}(1)}$. The frequency of an atom, ν_0, located at point 1, when seen by an observer at point 2 is, hence, given by

$$\nu = \nu_0 \sqrt{\frac{g_{00}(1)}{g_{00}(2)}}. \qquad (3.77)$$

For a gravitational field like that of Schwarzschild, one therefore obtains for the frequency shift per unit frequency:

$$\frac{\Delta\nu}{\nu_0} = \frac{\nu - \nu_0}{\nu_0} \approx -\frac{Gm}{c^2}\left(\frac{1}{r_1} - \frac{1}{r_2}\right), \qquad (3.78)$$

to first order in Gm/c^2r. If we take r_1 to be the observed radius of the Sun and r_2 the radius of the Earth's orbit around the Sun (thus neglecting completely the Earth's gravitational field), then $\Delta\nu/\nu_0 = -2.12 \times 10^{-6}$. This frequency shift is usually referred to as the *gravitational red shift*.

The gravitational red shift was tested for the Sun and for white dwarfs, and it was suggested that it be tested by atomic clocks. The red shift was also observed directly using the Mössbauer effect by Pound and Rabka, and by Cranshaw, Schiffer and Whitehead. The latter employed Fe^{57} and a total height difference of 12.5 meters. A red shift 0.96±0.45 times the predicted value was observed by them. Pound and Rabka's result is more precise. They obtained a red shift 1.05±0.10 times the predicted value.

3.6.2 Effects on planetary motion

One assumes that test particles move along geodesics in the gravitational field (see next section), and that planets have small masses as compared with the mass of the Sun, thus behaving like test particles. Consequently, to find the equation of motion of a planet moving in the gravitational field of the Sun one has to write the geodesic equation in the Schwarzschild field. In fact one does not need the exact solution (3.76) but only its first approximation,

$$g_{00} = 1 - 2Gm/r,$$
$$g_{0r} = 0,$$
$$g_{rs} = -\delta_{rs} - 2Gmx^r x^s/r^3. \tag{3.79}$$

In the above equations the speed of light is taken as unity.

Using the approximate metric (3.79) in the geodesic equation (3.43) gives

$$\ddot{\mathbf{x}} - Gm\nabla\frac{1}{r}$$

$$= Gm\left\{2(\dot{\mathbf{x}}^2)\nabla\frac{1}{r} - 2Gm\frac{1}{r}\nabla\frac{1}{r} - 2\left(\dot{\mathbf{x}}\cdot\nabla\frac{1}{r}\right)\dot{\mathbf{x}} + \frac{3}{r^5}(\mathbf{x}\cdot\dot{\mathbf{x}})^2\mathbf{x}\right\}, \tag{3.80}$$

where we have used three-dimensional notation and a dot denotes differentiation with respect to t. Multiplying Eq. (3.80), vectorially, by the radius vector $\mathbf{x}$ gives

$$\mathbf{x} \times \ddot{\mathbf{x}} = -2Gm\left(\dot{\mathbf{x}}\cdot\nabla\frac{1}{r}\right)(\mathbf{x}\times\dot{\mathbf{x}}), \tag{3.81}$$

thus leading to the first integral

$$\mathbf{x} \times \dot{\mathbf{x}} = \mathbf{J}e^{-2Gm/r}, \tag{3.82}$$

where $\mathbf{J}$ is a *constant* vector, the angular momentum per mass unit.

Hence the radius vector $\mathbf{x}$ moves in a plane perpendicular to the vector $\mathbf{J}$, as in Newtonian mechanics. Introducing in this plane polar coordinates r, ϕ to describe the motion of the planet, the equation of motion (3.80), consequently, decomposes into

$$\ddot{r} - r\dot{\phi}^2 + \frac{Gm}{r^2} = \frac{Gm}{r^2}\left\{3\dot{r}^2 - 2r^2\dot{\phi}^2 + 2\frac{Gm}{r}\right\}, \tag{3.83a}$$

$$r^2\dot{\phi} = Je^{-2Gm/r}, \tag{3.83b}$$

where J is the magnitude of the vector $\mathbf{J}$.

Introducing now the new variable $u = 1/r$ one can rewrite Eqs. (3.83) in terms of $u(\phi)$:

$$u'' + u - \frac{Gm}{J^2} = Gm\left(-u'^2 + 2u^2 + 2\frac{Gm}{J^2}u\right). \qquad (3.84)$$

Here a prime denotes a derivation with respect to the angle ϕ.

Let us try a solution of the form

$$u = b\left(1 + \epsilon \cos \alpha\phi\right). \qquad (3.85)$$

Here ϵ is the eccentricity and α is some parameter to be determined and whose value in the usual nonrelativistic mechanics is unity. The other constant b is related to J in the nonrelativistic mechanics by $Gm/J^2 = b$. Using the above solution in Eq. (3.84) and equating coefficients of $\cos \alpha\phi$ gives

$$\alpha^2 = 1 - 2Gm\left(2b + Gm/J^2\right). \qquad (3.86)$$

Substituting for Gm/J^2 its nonrelativistic value b then gives $\alpha^2 = 1 - 6Gmb$, or, to a first approximation in Gm,

$$\alpha = 1 - 3Gmb. \qquad (3.87)$$

Successive perihelia occur when

$$(1 - 3Gmb)(2\pi + \Delta\phi) = 2\pi. \qquad (3.88)$$

Consequently, there will be an advance in the perihelion of the orbit per revolution given by $\Delta\phi = 6\pi Gmb$, or $\Delta\phi = 6\pi Gm/a(1 - \epsilon^2)$ if we make use of the nonrelativistic value of the constant b, and where a is the semimajor axis of the orbit. Reinstating now c, the velocity of light, finally gives for the perihelion advance

$$\Delta\phi = \frac{6\pi Gm}{c^2 a(1 - \epsilon^2)}, \qquad (3.89)$$

in radians per revolution (see Figs. 3.1 and 3.2).

We list below the calculated values of $\Delta\phi$ per century for four planets:

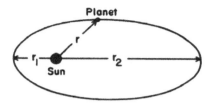

Figure 3.1: Newtonian limit of planetary motion. The motion is described by a closed ellipse if the effect of other planets is completely neglected.

Figure 3.2: Planetary elliptic orbit with perihelion advance. The effect is a general relativistic one. The advance of the perihelion is given by $\Delta\phi$ in radians per revolution, where $\Delta\phi = 6\pi Gm/c^2 a(1 - \epsilon^2)$, with m being the mass of the Sun, a the semimajor axis, and ϵ the eccentricity of the orbit of the planet.

Planet	$\Delta\phi$
Mercury	43.03″
Venus	8.60″
Earth	3.80″
Mars	1.35″

The astronomical observations for the planet Mercury give 43.11 ± 0.45 sec per century, in good agreement with the calculated value.

3.6.3   Deflection of light

To discuss the deflection of light in the gravitational field we must again solve the geodesic equation, but now with the null condition $ds = 0$. Using the appropriate solution (3.79) then gives for $g_{\mu\nu}dx^\mu dx^\nu = 0$

$$\left(1 + \frac{2Gm}{r}\right)\left[\dot{\mathbf{x}} \cdot \dot{\mathbf{x}} + \frac{2Gm}{r^3}(\mathbf{x} \cdot \dot{\mathbf{x}})^2\right] = 1. \tag{3.90}$$

Using polar coordinates r, ϕ, consequently, gives to the first approximation in Gm

$$\dot{r}^2 + r^2\dot{\phi}^2 + \frac{4Gm\dot{r}^2}{r} + 2Gmr\dot{\phi}^2 = 1. \tag{3.91}$$

Again changing variables into $u(\phi) = 1/r$, and using Eq. (3.83b), gives

$$u'^2 + u^2 + 2Gmu\left(2u'^2 + u^2\right) = J^{-2}e^{4Gmu}. \tag{3.92}$$

Differentiation of this equation with respect to ϕ gives

$$u'' + u + Gm\left(2u'^2 + 4uu'' + 3u^2\right) = 2GmJ^{-2}, \tag{3.93}$$

to the first approximation in Gm.

To solve Eq. (3.93) we note that in the lowest approximation one has

$$u'^2 \approx J^{-2} - u^2, \tag{3.94}$$

$$u'' \approx -u. \tag{3.95}$$

Using these values in Eq. (3.93) gives

$$u'' + u = 3Gmu^2, \tag{3.96}$$

for the orbit of the light ray. In the lowest approximation u satisfies $u'' + u = 0$, whose solution is a straight line (see Fig. 3.3)

$$u = \frac{\cos \phi}{R}, \tag{3.97}$$

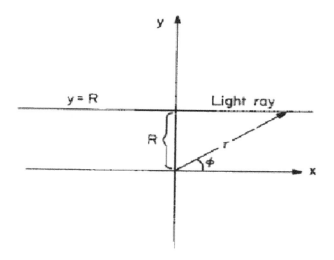

Figure 3.3: Light ray when the effect of the central body's grav-itational field is completely neglected. The light ray then moves along the straight line $y = r \sin \phi = R$ =constant, namely, $u = 1/r = (1/R) \sin \phi$.

where R is a constant. This shows that $r = 1/u$ has a minimum value R at $\phi = 0$. Substituting into the right-hand side of Eq. (3.96) then gives

$$u'' + u = 3\frac{Gm}{R^2} \cos^2 \phi. \qquad (3.98)$$

The solution of this equation is

$$u = \frac{\cos \phi}{R} + \frac{Gm}{R^2}\left(1 + \sin^2 \phi\right). \qquad (3.99)$$

Introducing now Cartesian coordinates $x = r \cos \phi$ and $y = r \sin \phi$, the above equation gives

$$x = R - \frac{Gm}{R}\frac{x^2 + 2y^2}{\sqrt{x^2 + y^2}}. \qquad (3.100)$$

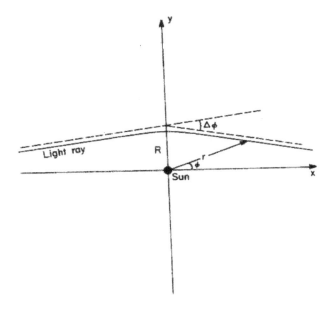

Figure 3.4: Bending of a light ray in the gravitational field of a spherically symmetric body. The angle of deflection $\Delta\phi = 4Gm/c^2R$, where m is the mass of the central body and R is the closest distance of the light ray from the center of the body.

For large values of $\mid y \mid$ this equation becomes

$$x \approx R - \frac{2Gm}{R} \mid y \mid. \tag{3.101}$$

Hence, asymptotically, the orbit of the light ray is a straight line in space. This result is expected, since far away from the central mass the space is flat. The angle $\Delta\phi$ between the two asymptotes is, however, equal to (see Fig. 3.4)

$$\Delta\phi = 4\frac{Gm}{c^2R}, \tag{3.102}$$

in units in which c is different from unity.

The angle $\Delta\phi$ represents the angle of *deflection* of a light ray in passing through the Schwarzschild field. For a light ray just grazing the Sun Eq. (3.102) gives $\Delta\phi = 1.75$ sec. Observations indeed confirm this result; one of the latest results gives 1.75 ± 0.10 sec.

3.6.4 Gravitational radiation experiments

Weber has developed methods to detect gravitational waves that Einstein's gravitational field equations predict. The experiment involves detectors at opposite ends of a 1000 km baseline. Sudden increases in detector output were observed by him roughly once in several days, coincident within the resolution time of 0.25 seconds.

Weber's apparatus measures the Fourier transform of the Riemann tensor. The method uses the fact that the distance η^μ between two neighboring test particles, which follow geodesics, satisfies the *geodesic deviation* equation

$$\frac{\delta^2 \eta^\mu}{\delta s^2} + R^\mu{}_{\alpha\nu\beta}\lambda^\alpha \eta^\nu \lambda^\beta = 0, \qquad (3.103)$$

where λ^α is the tangent vector to one of the geodesics, and $\delta/\delta s = \lambda^\alpha \nabla_\alpha$ is a directional covariant derivative. Weber measured the strains of a large aluminum cylinder, having mass of the order 10^6 grams, by means of a piezoelectric crystal attached to the cylinder which transforms the mechanical movement into an electric current. The detector was developed for operation in the vicinity of 1662 cycle/sec. A high frequency source was developed for dynamic gravitational fields and the detector was tested by doing a communication experiment with high frequency Coulomb fields. New methods, LEIGO, are these days operating.

3.6.5 Radar experiment

Shapiro has designed a radar experiment to test general relativity by measuring the effect of solar gravity on time delays of round-trip travel times of radar pulses transmitted from the Earth toward an inner planet, i.e., Venus or Mercury. The experiment is based on the phenomenon that electromagnetic waves "slow down" in the gravitational field. Within the framework of general relativity there should be an anomalous delay of 200 microseconds in the arrival time of a radar echo from Mercury, positioned on the far side of the Sun near the limb.

For example, if we calculate the proper time τ at $r = r_2$ for a radial round-trip travel $r_2 \to r_1 \to r_2$, with $r_2 > r_1$, of a radar pulse in the Schwarzschild field, and subtract from τ the corresponding value τ_0 when the spherical mass $m = 0$, we find

$$\Delta\tau = \frac{4Gm}{c^3}\left(\ln\frac{r_2}{r_1} - \frac{r_2 - r_1}{r_2}\right) + O\left(m^2\right). \tag{3.104}$$

In general one finds

$$\Delta\tau \approx \frac{4Gm}{c^3}\left(\ln\frac{r_e + r_p + R}{r_e + r_p - R}\right), \tag{3.105}$$

where r_e is the Earth-Sun distance, r_p the planet-Sun distance, and R the Earth-planet distance.

Shapiro found that the retardation of radar signals are 1.02 ± 0.05 times the corresponding effect predicted by general relativity.

3.6.6 Low-temperature experiments

Schiff has proposed an experiment to check the equations of motion in general relativity by means of a gyroscope, which is forced to go around the Earth either in a stationary laboratory fixed to the Earth or a satellite. The unique experiment is made possible by complete use of a low-temperature environment, and the properties of superconductors, including the use of zero magnetic fields and ultrasensitive magnetometry. Schiff has calculated, using results obtained by Papapetrou for the motion of spinning bodies in general relativity, that a perfect gyroscope subject to no torques will experience an anomalous precession with respect to the fixed stars as it travels around the Earth.

3.7 Equations of motion

3.7.1 Geodesic postulate

In the last section it was assumed that the planet's motion around the Sun is described by the geodesic equation (3.43). The assumption that the equations of motion of a test particle, moving in gravitational field, are given by the geodesic equation is known as the *geodesic postulate* and was suggested by Einstein in his first article on the general theory of relativity.

Eleven years later when Einstein and Grommer showed that the geodesic postulate need not be assumed, but that it rather follows from the gravitational field equations; this is a consequence of the nonlinearity of the field equations along with the fact that they satisfy the four contracted Bianchi identities (see Section 3.1). The discovery of Einstein and Grommer is considered to be one of the most important achievements, and one of the most attractive features of the general theory of relativity. Later on Infeld and Schild showed that the equations of motion of a test particle are given by the geodesic equation in the *external gravitational* field. This result, however, does not differ from the geodesic postulate because, by definition, a test particle has no self-field.

3.7.2 Equations of motion as a consequence of field equations

In order to establish the relation between the Einstein field equations and the equations of motion one proceeds as follows. We have seen in Section 3.4 that because of the contracted

Bianchi identity it follows that the energy-momentum tensor $T^{\mu\nu}$ satisfies a generally covariant conservation law of the form given by Eq. (3.51). Consequently, one obtains for the energy-momentum tensor *density* $\mathcal{T}^{\mu\nu}$

$$\nabla_\nu \mathcal{T}^{\mu\nu} = \partial_\nu \mathcal{T}^{\mu\nu} + \Gamma^\mu_{\alpha\beta} \mathcal{T}^{\alpha\beta} = 0, \qquad (3.106)$$

where $\mathcal{T}^{\mu\nu} \equiv \sqrt{-g} T^{\mu\nu}$.

For a system of N particles of finite masses, represented as singularities of the gravitational field, $\mathcal{T}^{\mu\nu}$ may be taken in the form

$$\mathcal{T}^{\mu\nu} = \sum_{A=1}^{N} m_A v_A^\mu v_A^\nu \delta_A \left(\mathbf{x} - \mathbf{z}_A\right). \qquad (3.107)$$

Here z_A^μ are the coordinates of the Ath particle. (Roman capital indices, $A, B, \cdots$, run from 1 to N. For these indices the summation convention will be suspended.) Also $v^\mu = \dot{z}^\mu = dz^\mu/dt$ $(v_A^0 = \dot{z}_A^0 = 1)$, and δ is the three-dimensional Dirac delta function satisfying the following conditions:

$$\delta\left(\mathbf{x}\right) = 0; \quad \text{for} \quad x \neq 0, \qquad (3.108\text{a})$$

$$\int \delta\left(\mathbf{x} - \mathbf{z}\right) d^3x = 1, \qquad (3.108\text{b})$$

$$\int f\left(\mathbf{x}\right) \delta\left(\mathbf{x} - \mathbf{z}\right) d^3x = f\left(\mathbf{z}\right), \qquad (3.108\text{c})$$

for any continuous function $f\left(\mathbf{x}\right)$ in the neighborhood of $\mathbf{z}$. In Eq. (3.107), m_A is a function of time which may be called the *inertial mass* of the Ath particle.

If we put the energy-momentum tensor density (3.107) into (3.106) and integrate over the three-dimensional region surrounding the first singularity, we obtain

$$\frac{dp^\mu}{dt} = \int F^\mu \delta\left(\mathbf{x} - \mathbf{z}\right) d^3x, \qquad (3.109)$$

where $p^\mu = mv^\mu$ and $F^\mu = -m\Gamma^\mu_{\alpha\beta} v^\alpha v^\beta$, and where we have put, for simplicity, $m = m_1$, $z^\mu = z_1^\mu$, $v^\mu = v_1^\mu$, and $\delta\left(\mathbf{x} - \mathbf{z}\right) = \delta_1\left(\mathbf{x} - \mathbf{z}_1\right)$.

3.7.3 Self-action terms

Equation (3.109) may be interpreted as an "exact equation of motion" of the first particle. However, since the Christoffel symbols are singular at the location of the particle, the equation contains infinite self-action terms. However, it was shown that these terms can be removed as follows.

Putting Eq. (3.107) into Eq. (3.106) we obtain

$$\partial_0 \left[\sum_{A=1}^{N} m_A v_A^\mu \delta_A \right] + \partial_n \left[\sum_{A=1}^{N} m_A v_A^\mu v_A^n \delta_A \right] + \sum_{A=1}^{N} m_A \Gamma_{\alpha\beta}^\mu v_A^\alpha v_A^\beta \delta_A = 0,$$
(3.110)

where Latin indices run from 1 to 3. The first term on the left-hand side of Eq. (3.106) can be written as

$$\partial_0 \left[\sum_{A=1}^{N} m_A v_A^\mu \delta_A \right] = \sum_{A=1}^{N} \partial_0 \left(m_A v_A^\mu \right) \delta_A + \sum_{A=1}^{N} m_A v_A^\mu \partial_0 \delta_A, \quad (3.111)$$

with

$$\partial_0 \delta_A = \partial_0 \delta_A \left(x^s - z_A^s \right) = -\partial_n \delta_A v_A^n. \tag{3.112}$$

Using the above results in Eq. (3.110), we obtain

$$\sum_{A=1}^{N} \left\{ \frac{d \left(m_A v_A^\mu \right)}{dt} + m_A \Gamma_{\alpha\beta}^\mu v_A^\alpha v_A^\beta \right\} \delta_A = 0. \tag{3.113}$$

Equation (3.113), which is identical with Eq. (3.106), is satisfied at any spacetime point, since otherwise the Bianchi identity or the Einstein field equations will not be satisfied.

We now examine the behavior of Eq. (3.113) in the infinitesimal neighborhood of the first singularity, which we assume not to contain any other singularity. In this region $\delta_B \left(\mathbf{x} - \mathbf{z}_B \right) = 0$ for $B = 2, 3, \cdots, N$. Hence Eq. (3.113) gives for the conservation law near the first singularity

$$\left\{ \frac{d \left(m v^\mu \right)}{dt} + m \Gamma_{\alpha\beta}^\mu v^\alpha v^\beta \right\} \delta \left(\mathbf{x} - \mathbf{z} \right) = 0. \tag{3.114}$$

Let us further assume that the Christoffel symbols near the first singularity can be expanded into a power series in the infinitesimal distance r, defined by $r^2 = (x^s - z^s)(x^s - z^s)$, where $z^s = z_1^s$, in the vicinity of the first particle. Then we have

$$\Gamma^\mu_{\alpha\beta} = {}_{-k}\Gamma^\mu_{\alpha\beta} + {}_{-k+1}\Gamma^\mu_{\alpha\beta} + \cdots + {}_0\Gamma^\mu_{\alpha\beta} + \cdots, \tag{3.115}$$

where the indices written in subscripts on the left of a function indicate its behavior with respect to r, and k is a positive integer.

For example ${}_0\Gamma^\mu_{\alpha\beta}$ is that part of the Christoffel symbol which varies as r^0, i.e., is finite at the location of the first particle. When one uses spherical coordinates r, θ and ϕ, one can write

$$-k\Gamma^\mu_{\alpha\beta} = r^{-k}A^\mu_{\alpha\beta}(\theta, \phi), \tag{3.116a}$$

$$-k+1\Gamma^\mu_{\alpha\beta} = r^{-k+1}B^\mu_{\alpha\beta}(\theta, \phi), \tag{3.116b}$$

$$\cdot \quad \cdot \quad \cdot \quad \cdot \quad \cdot \quad \cdot \quad \cdot$$

$$_0\Gamma^\mu_{\alpha\beta} = D^\mu_{\alpha\beta}(\theta, \phi), \quad \text{etc.} \tag{3.116c}$$

Terms like ${}_1\Gamma^\mu_{\alpha\beta}$, ${}_2\Gamma^\mu_{\alpha\beta}$, etc., however, need not be taken into account when one puts the above expansion into Eq. (3.114) since $r^j\delta(\mathrm{x} - \mathrm{z}) = 0$ for any positive integer j. If we denote now $mA^\mu_{\alpha\beta}v^\alpha v^\beta, \cdots$ by $A^\mu, \cdots$ we can write Eq. (3.114) in the form

$$\left\{ r^{-k}A^\mu + r^{-k+1}B^\mu + \cdots + r^{-1}C^\mu + D_1^\mu \right\}\delta(\mathrm{x} - \mathrm{z}) = 0, \tag{3.117}$$

where we have used the notation $D_1^\mu = d(mv^\mu)/dt + D^\mu$.

In order to get rid of terms proportional to negative powers of r in Eq. (3.117) we proceed as follows. Multiplying Eq. (3.117) by r^k and using $r^j\delta(\mathrm{x} - \mathrm{z}) = 0$ we obtain

$$A^\mu(\theta, \phi)\delta(\mathrm{r}) = 0, \tag{3.118}$$

the integration of which over the three-dimensional region yields, using spherical coordinates,

$$\int\int A^\mu(\theta, \phi)\sin\theta d\theta d\phi \int r^2\delta(\mathrm{r}) = 0. \tag{3.119}$$

From the property of the delta-function

$$\int \delta(\mathbf{r}) \, d^3x = \int\int \sin\theta d\theta d\phi \int \delta(\mathbf{r}) \, r^2 dr = 1, \qquad (3.120)$$

one obtains $\int \delta(\mathbf{r}) \, r^2 dr = (4\pi)^{-1}$. Hence we obtain

$$\int\int A^\mu(\theta, \phi) \sin\theta d\theta d\phi = 0, \qquad (3.121)$$

independent of the value of the variable R. Thus the angular distribution of $A^\mu(\theta, \phi)$ is such that its average equals zero.

However, not only does the above equation hold, but also (s is any finite positive integer)

$$a(r) = r^{-s} \int\int A^\mu(\theta, \phi) \sin\theta d\theta d\phi = 0, \qquad (3.122)$$

for small values of r as well as when r tends to zero, as can be verified by using L'Hospital's theorem, for example. It follows then that $a(r)$ is a function of r whose value is zero for any small r, including $r = 0$. Using the property of delta-function we obtain

$$\int r^2 \delta(\mathbf{r}) \, dr = (4\pi)^{-1} f(0), \qquad (3.123)$$

for any continuous function of r. Since $a(r)$ is certainly continuous, one obtains

$$\int r^2 \delta(\mathbf{r}) \, a(r) \, dr = 0. \qquad (3.124)$$

Hence when one integrates Eq. (3.117) over the three-dimensional space, there will be no contribution from the first term.

In order to show that the second term of Eq. (3.117) will also not contribute to the three-dimensional integration of the same equation, we multiply it by r^{k-1}. We obtain now, after neglecting terms that do not contribute,

$$\left\{ r^{-1} A^\mu(\theta, \phi) + B^\mu(\theta, \phi) \right\} \delta(\mathbf{r}) = 0. \qquad (3.125)$$

Integration of this equation, again using spherical coordinates, shows that the first term will not contribute anything because of Eq. (3.124), and we are left with

$$\int \int B^\mu (\theta, \phi) \sin\theta d\theta d\phi \int r^2 \delta (\mathbf{r}) \, dr = 0. \tag{3.126}$$

Hence we have

$$\int \int B^\mu (\theta, \phi) \sin\theta d\theta d\phi = 0, \tag{3.127}$$

independent of r. From this equation one obtains another one, analogous to Eq. (3.124) but with B^μ instead of A^μ:

$$\int r^2 \delta (\mathbf{r}) \, b (r) \, dr = 0, \tag{3.128}$$

with

$$b (r) = r^{-s} \int \int B^\mu (\theta, \phi) \sin\theta d\theta d\phi = 0. \tag{3.129}$$

Proceeding in this way, one verifies that the angular distribution of all functions A^μ, B^μ, etc., is such that they all satisfy equations like Eqs. (3.121) and (3.127). Hence it is clear that one obtains

$$\int D_1^\mu (\theta, \phi) \, \delta (\mathbf{r}) \, d^3x = 0, \tag{3.130}$$

which gives

$$\frac{dp^\mu}{dt} + m v^\alpha v^\beta \int {}_0\Gamma^\mu_{\alpha\beta} \delta (\mathbf{r}) \, d^3x = 0, \tag{3.131}$$

or equivalently

$$\dot{v}^k + v^\alpha v^\beta \int \left({}_0\Gamma^k_{\alpha\beta} - v^k {}_0\Gamma^0_{\alpha\beta} \right) \delta (\mathbf{r}) \, d^3x = 0. \tag{3.132}$$

Equation (3.132) is the "exact equation of motion".

3.7.4   Einstein-Infeld-Hoffmann method

Having found the law of motion (3.131), one can now proceed to find the equation of motion of two finite masses, each moving in the field produced by both of them. In the following we find such an equation of motion in the case for which the particles' velocities are much smaller than the speed of light. Moreover, we will confine ourselves to an accuracy of post-Newtonian. This means the equation of motion obtained will contain the Newtonian equation as a limit, but is a first generalization of it. Such an equation was first obtained by Einstein, Infeld, and Hoffmann. To obtain this equation we solve the field equations and formulate the equations of motion explicitly by means of an approximation method, the Einstein-Infeld-Hoffmann (EIH) method, to be described below.

Let us assume a function ϕ developed in a power series in the parameter $\lambda = 1/c$, where c is the speed of light. One then has

$$\phi = {}_0\phi + {}_1\phi + {}_2\phi + \cdots. \tag{3.133}$$

The indices written as left subscripts indicate the order of λ absorbed by the ϕ's.

If a function $\phi(x)$ varies rapidly in space but slowly with x^0, then we are justified in not treating all its derivatives in the same manner. The derivatives with respect to x^0 will be of a higher order than the space derivatives. We thus write

$$\partial_0 ({}_l\psi) = {}_{l+1}\psi. \tag{3.134}$$

That is, differentiation with respect to x^0 raises the order by one. Thus if the coordinates z^s of a particle are considered to be of order zero, $\dot{z}^s$ will be of order one, and $\ddot{z}^s$ of order two. Using now the Newtonian approximation mass $\times$ acceleration=mass $\times$ mass/(distance)^2, we see the mass is of order two. In all the power developments we take into account only even or only odd powers of $1/c$. (The expansion of the metric tensor, etc., in a power series in c^{-2} (such as $\phi = {}_0\phi + {}_2\phi + \cdots$, or $\phi = {}_1\phi + {}_3\phi + \cdots$) corresponds to the choice of the symmetric Green function, thus excluding radiation.)

Thus, because of the order with which we start m and $\dot{z}^s$, we have

$$\mathcal{T}^{00} = {}_2\,\mathcal{T}^{00} + {}_4\,\mathcal{T}^{00} + \cdots,$$

$$\mathcal{T}^{0n} = {}_3\,\mathcal{T}^{0n} + {}_5\,\mathcal{T}^{0n} + \cdots, \tag{3.135}$$

$$\mathcal{T}^{mn} = {}_4\,\mathcal{T}^{mn} + {}_6\,\mathcal{T}^{mn} + \cdots.$$

As to the metric tensor, we write

$$g_{\mu\nu} = \eta_{\mu\nu} + h_{\mu\nu}, \qquad g^{\mu\nu} = \eta^{\mu\nu} + h^{\mu\nu}. \tag{3.136}$$

The gravitational field equations can be written as

$$\sqrt{-g}\,R_{\alpha\beta} = \kappa\left(\mathcal{T}_{\alpha\beta} - \frac{1}{2}g_{\alpha\beta}\mathcal{T}\right), \tag{3.137}$$

where $\mathcal{T} = \mathcal{T}_{\mu\nu}g^{\mu\nu}$, and $R_{\alpha\beta}$ is the Ricci tensor. From the right-hand side of the field equations it follows that R_{00} and R_{mn} (when $m = n$) start with order two, R_{mn} (when $m \neq n$) start with order four, while R_{0m} starts with order three. The lowest order expressions of the left-hand side are

$$R_{00} \approx \frac{1}{2}h_{00,ss},$$

$$R_{0m} \approx \frac{1}{2}\left(h_{0m,ss} - h_{0s,ms} - h_{ms,0s} + h_{ss,0m}\right), \tag{3.138}$$

$$R_{mn} \approx \frac{1}{2}\left(h_{mn,ss} - h_{ms,ns} - h_{ns,ms} - h_{00,mn} + h_{ss,mn}\right),$$

where a comma denotes a partial derivative, $\phi_{,s} = \partial_s\phi$. Hence we have

$$h_{00} = {}_2h_{00} + {}_4h_{00} + \cdots,$$

$$h_{0m} = {}_3h_{0n} + {}_5h_{0n} + \cdots, \tag{3.139}$$

$$h_{mn} = {}_2h_{mn} + {}_4h_{mn} + \cdots.$$

3.7.5 Newtonian equation of motion

We now find the equation of motion in the lowest (Newtonian) approximation. We do it in such a way as to make the generalization to the post-Newtonian approximation as simple as possible.

Because of Eqs. (3.137) and (3.138), the field equations of the lowest order are in h_{00},

$$\frac{1}{2}{}_2h_{00,ss} = \kappa \left({}_2T^{00} - \frac{1}{2}{}_2T^{00} \right) = \frac{\kappa}{2}{}_2T^{00} = \frac{\kappa}{2}\sum_{A=1}^{2} \mu_A \delta_A, \quad (3.140)$$

where, for simplicity, we have put $\mu_A = {}_2m_A$. Hence the equation obtained is

$$_2h_{00,ss} = \kappa \sum_{A=1}^{2} \mu_A \delta_A. \quad (3.141)$$

The solution of this equation that represents two masses is

$$_2h_{00} = -2G \sum_{A=1}^{2} \mu_A r_A^{-1}, \quad (3.142)$$

where $r_A^2 = (x^s - z_A^s)(x^s - z_A^s)$. Using $_2h_{00}$ in the equation of motion (3.132), we obtain in the lowest (second) order for the equation of motion of the first particle

$$\ddot{z}_1^k - G \int \partial_k \left(\mu_2 r_2^{-1} \right) \delta \left(\mathbf{x} - \mathbf{z}_1 \right) d^3x = 0. \quad (3.143)$$

This gives

$$\ddot{z}_1^k = G \frac{\partial}{\partial z_1^k} \frac{\mu_2}{z}, \quad (3.144)$$

where $z^2 = (z_1^s - z_2^s)(z_1^s - z_2^s)$. Equation (3.144) is, of course, the Newtonian equation of motion.

3.7.6   Einstein-Infeld-Hoffmann equation

To find the equation of motion up to the fourth order, we must know besides $_2h_{00}$ the functions $_4h_{00}$, $_3h_{0n}$ and $_2h_{mn}$. The second and third functions are easy to find. The left-hand side of the corresponding equations is written out in Eq. (3.138), whereas the right-hand side is given by Eq. (3.137) and it is $-\kappa \sum \mu_A \dot{z}_A^m \delta_A$ for the $0m$ component, and $\frac{\kappa}{2}\delta_{mn} \sum \mu_A \delta_A$ for the mn component. Therefore, for the $_2h_{mn}$ we have the equation

$$_2h_{mn,ss} - {_2h_{ms,ns}} - {_2h_{ns,ms}} + {_2h_{ss,mn}} - {_2h_{00,mn}} = \delta_{mn2}h_{00,ss},$$
(3.145)

whose solution is

$$_2h_{mn} = \delta_{mn}\,{_2h_{00}}.$$
(3.146)

The equation for $_3h_{0n}$ is

$$_3h_{0n,ss} - {_3h_{0s,ns}} - {_2h_{ns,0s}} + {_2h_{ss,n0}} = -2\kappa \sum \mu_A \dot{z}_A^n \delta_A.$$
(3.147)

Using the value of $_2h_{mn}$ in terms of the $_2h_{00}$ found above, we obtain

$$_3h_{0n,ss} - {_3h_{0s,ns}} + 2_2h_{00,n0} = -2\kappa \sum_{A=1}^{2} \mu_A \dot{z}_A^n \delta_A.$$
(3.148)

The solution of this equation is

$$_3h_{0n} = 4G \sum_{A=1}^{2} \mu_A \dot{z}_A^n r_A^{-1}.$$
(3.149)

Calculation of $_4h_{00}$ is somewhat more complicated. The relevant part of $_4h_{00}$, for two masses, that contributes to the equation of motion of the first particle, is

$$_4h_{00} \approx G\left\{2G\mu_2^2 r_2^{-2} - 3\mu_2 \dot{z}_2^s \dot{z}_2^s r_2^{-1} - \mu_2 r_{2,00} + 2G\mu_1\mu_2 \left(zr_2\right)^{-1}\right\}.$$
(3.150)

Using these values for $_4h_{00}$, $_3h_{0n}$, and $_2h_{mn}$ in the equation of motion (3.132) gives, for the two-body problem:

$$\ddot{z}_1^n - \mu_2 \frac{\partial\left(1/z\right)}{\partial z_1^n}$$

$$= \mu_2\Big\{\Big(\dot{z}_1^s\dot{z}_1^s + \frac{3}{2}\dot{z}_2^s\dot{z}_2^s - 4\dot{z}_1^s\dot{z}_2^s - 4\frac{\mu_2}{z} - 5\frac{\mu_1}{z}\Big)\frac{\partial\left(1/z\right)}{\partial z_1^n}$$

$$+ \big[4\dot{z}_1^s\left(\dot{z}_2^n - \dot{z}_1^n\right) + 3\dot{z}_1^n\dot{z}_2^s - 4\dot{z}_2^n\dot{z}_2^s\big]\frac{\partial\left(1/z\right)}{\partial z_1^s} + \frac{1}{2}\dot{z}_2^s\dot{z}_2^r\frac{\partial^3 z}{\partial z_1^s\partial z_1^r\partial z_1^n}\Big\}.$$

$$(3.151)$$

In Eq. (3.151) the Newtonian gravitational constant G was taken as equal to 1. The equation of motion for the second particle is obtained by replacing μ_1, μ_2, z_1, z_2 by μ_2, μ_1, z_2, z_1, respectively.

Equation (3.151) is known as the Einstein-Infeld-Hoffmann equation of motion, and is a generalization of the Newton equation. The essential relativistic correction may be obtained by fixing one of the particles. Writing M for μ_2, neglecting μ_1 and $\dot{z}_2^s$, and using an obvious three-dimensional vector notation, Eq. (3.151) simplifies to

$$\ddot{\mathbf{z}} - M\nabla\left(\frac{1}{z}\right) = M\left\{\left(\dot{\mathbf{z}}\cdot\dot{\mathbf{z}} - \frac{4M}{z}\right)\nabla\left(\frac{1}{z}\right) - 4\dot{\mathbf{z}}\left(\dot{\mathbf{z}}\cdot\nabla\frac{1}{z}\right)\right\},$$

$$(3.152)$$

where $\mathbf{z}$ denotes the three-vector z_1^s.

3.8 Decomposition of the Riemann tensor

The Riemann curvature tensor $R_{\alpha\beta\gamma\delta}$ can be decomposed into its *irreducible components*. These are the Weyl conformal tensor $C_{\alpha\beta\gamma\delta}$, the tracefree Ricci tensor $S_{\alpha\beta}$, and the Ricci scalar curvature R. The tensor $S_{\alpha\beta}$ is defined by

$$S_{\alpha\beta} = R_{\alpha\beta} - \frac{1}{4}g_{\alpha\beta}R, \tag{3.153}$$

where $R_{\alpha\beta}$ is the ordinary Ricci tensor.

The decomposition can be written symbolically as

$$R_{\alpha\beta\gamma\delta} = C_{\alpha\beta\gamma\delta} \bigoplus S_{\alpha\beta} \bigoplus R. \tag{3.154}$$

No new quantities can be obtained from any of the above three irreducible components by contraction of their indices.

When written in full details, the decomposition (3.154) will then have the form:

$$R_{\rho\sigma\mu\nu} = C_{\rho\sigma\mu\nu} + \frac{1}{2}\left(g_{\rho\mu}S_{\sigma\nu} - g_{\rho\nu}S_{\sigma\mu} - g_{\sigma\mu}S_{\rho\nu} + g_{\sigma\nu}S_{\rho\mu}\right)$$

$$-\frac{1}{12}\left(g_{\rho\nu}g_{\sigma\mu} - g_{\rho\mu}g_{\sigma\nu}\right)R. \tag{3.155}$$

It can also be written in the form:

$$R_{\rho\sigma\mu\nu} = C_{\rho\sigma\mu\nu} + \frac{1}{2}\left(g_{\rho\mu}R_{\sigma\nu} - g_{\rho\nu}R_{\sigma\mu} - g_{\sigma\mu}R_{\rho\nu} + g_{\sigma\nu}R_{\rho\mu}\right)$$

$$+\frac{1}{6}\left(g_{\rho\nu}g_{\sigma\mu} - g_{\rho\mu}g_{\sigma\nu}\right)R. \tag{3.156}$$

References

M. Carmeli, *Classical Fields: General Relativity and Gauge Theory*, John Wiley, New York, 1982 (reprinted by World Scientific, 2001).

D.J. Struik, *Lectures on Classical Differential Geometry*, Addison Wesley Press, Cambridge, 1950.

CHAPTER 4

COSMOLOGICAL GENERAL RELATIVITY

4.1 Introduction

In this chapter we will develop the theory of cosmological general relativity (CGR).

This theory is different from that of Einstein's general relativity theory discussed in the last chapter. While Einstein's theory deals with the gravitational field in the traditional form like electromagnetism, namely development in time of physical quantities, cosmological general relativity considers the Universe at one instant of time. It is like taking the picture of the Universe with a photographic camera and examines the distribution of galaxies in space, each with its own receding velocity. This is like in the theory of thermodynamics where one deals with the temperature and the pressure at some time and the relationship between them given by the equation of state. The collection of the locations of galaxies and their velocities provides a continuum of space and velocities. Therefore, instead of space and time in Einstein's theory we here have space and velocity, both of cases are in four dimensions.

The Einstein field equations

$$G_{\mu\nu} = R_{\mu\nu} - \frac{1}{2}g_{\mu\nu}R = \kappa T_{\mu\nu}, \qquad (4.1)$$

will be used in the space of distances and velocities.

The geodesic equation as the equation of motion

$$\frac{d^2 x^\rho}{ds^2} + \Gamma^\rho_{\alpha\beta} \frac{dx^\alpha}{ds} \frac{dx^\beta}{ds} = 0, \qquad (4.2)$$

will also be used in cosmological general relativity.

4.2   Space and velocity

Astronomers measure distances to faraway galaxies and their receding velocities. They do that in order to determine the expansion rate of the Universe. In Chapter 2 the foundations of the theory of the expansion of the Universe were given with the assumption that gravity is negligible. In this chapter we present the theory. We will derive a formula for the distance of the galaxy as a function of its velocity. The formula is very simple: $r(v) = (c\tau/\beta) \sinh \beta v/c$, where c is the speed of light in vacuum, τ is the Big Bang time ($\tau = 13.56 \pm 0.48 \text{Gyr}$), $\beta = \sqrt{1 - \Omega_m}$ and Ω_m is the mass density of the Universe. For $\Omega_m < 1$ this formula clearly shows that the Universe is expanding with acceleration, as experiments clearly show.

4.3   Gravitational field equations

In the four-dimensional spacevelocity, the most spherically symmetric metric is given by

$$ds^2 = \tau^2 dv^2 - e^\mu dr^2 - R^2 \left(d\theta^2 + \sin^2 \theta d\phi^2 \right), \qquad (4.3)$$

where μ and R are functions of v and r alone, and comoving coordinates $x^\mu = (x^0, x^1, x^2, x^3) = (\tau v, r, \theta, \phi)$ have been used. With the above choice of coordinates, the zero-component of the radial geodesic equation becomes an identity, and since r, θ and ϕ are constants along the geodesics, one has $dx^0 = ds$ and therefore $u^\alpha = u_\alpha = (1, 0, 0, 0)$. The metric (4.3) shows that the area of the sphere r =constant is given by $4\pi R^2$ and that R should satisfy $R' = \partial R/\partial r > 0$. The possibility that $R' = 0$ at a point r_0 is excluded since it would allow the lines r =constants at the neighboring points r_0 and $r_0 + dr$ to coincide at r_0, thus creating a caustic surface at which the comoving coordinates break down.

4.3.1 Universe expansion

As has been shown in Chapter 2 the Universe expansion at a given moment of time can be described by the null condition $ds = 0$, and if the expansion is spherically symmetric one has $d\theta = d\phi = 0$. The metric (4.3) then yields $\tau^2 dv^2 - e^\mu dr^2 = 0$, thus

$$\frac{dr}{dv} = \tau e^{-\mu/2}. \tag{4.4}$$

This is the differential equation that determines the Universe expansion. In the following we solve the gravitational field equations in order to find out the function $\mu(r.v)$.

The gravitational field equations, written in the form

$$R_{\mu\nu} = \kappa \left(T_{\mu\nu} - g_{\mu\nu} T/2 \right), \tag{4.5}$$

where

$$T_{\mu\nu} = \rho_{eff} u_\mu u_\nu + p \left(u_\mu u_\nu - g_{\mu\nu} \right), \tag{4.6}$$

with $\rho_{eff} = \rho - \rho_c$ and $T = T_{\mu\nu} g^{\mu\nu}$, are now solved. In the above equations $\kappa = 8\pi G/c^2 \tau^2$, ρ_c is the critical mass density which in this theory is a constant $\rho_c = 3h^2/8\pi G$, where $h = 1/\tau$. We use ρ_{eff} instead of just ρ since it allows us to have zero on the right-hand side of the Einstein field equations. Also, p is the pressure.

4.3.2 Energy-momentum tensor

One finds that the only nonvanishing components of $T_{\mu\nu}$ are $T_{00} = \tau^2 \rho_{eff}$, $T_{11} = c^{-1} \tau p e^\mu$, $T_{22} = c^{-1} \tau p R^2$ and $T_{33} = c^{-1} \tau p R^2 \sin^2 \theta$, and that $T = \tau^2 \rho_{eff} - 3c^{-1} \tau p$.

4.3.3 Independent field equations

One obtains three independent field equations (dot and prime denote derivatives with respect to the coordinates v and r)

$$e^{\mu}\left(2R\ddot{R} + \dot{R}^2 + 1\right) - R'^2 = -\kappa\tau c^{-1}e^{\mu}R^2 p, \qquad (4.7)$$

$$2\dot{R}' - R'\dot{\mu} = 0, \qquad (4.8)$$

$$e^{-\mu}\left[\frac{1}{R}R'\mu' - \left(\frac{R'}{R}\right)^2 - \frac{2}{R}R''\right] + \frac{1}{R}\dot{R}\dot{\mu} + \left(\frac{\dot{R}}{R}\right)^2 + \frac{1}{R^2} = \kappa\tau^2\rho_{eff}. \qquad (4.9)$$

4.4 Solution of the field equations

The solution of Eq. (4.8) satisfying the condition $R' > 0$ is given by

$$e^{\mu} = R'^2/\left(1 + f\left(r\right)\right), \qquad (4.10)$$

where $f\left(r\right)$ is an arbitrary function of the coordinate r and satisfies the condition $f\left(r\right) + 1 > 0$. Substituting (4.10) in the other two field equations (4.7) and (4.9) then gives

$$2R\ddot{R} + \dot{R}^2 - f = -\kappa c^{-1}\tau R^2 p, \qquad (4.11)$$

$$\frac{1}{RR'}\left(2\dot{R}\dot{R}' - f'\right) + \frac{1}{R^2}\left(\dot{R}^2 - f\right) = \kappa\tau^2\rho_{eff}, \qquad (4.12)$$

respectively.

4.4.1 Simple solution

The simplest solution of the above two equations, which satisfies the condition $R' = 1 > 0$, is given by $R = r$. Using this in Eqs. (4.11) and (4.12) gives $f\left(r\right) = \kappa c^{-1}\tau p r^2$, and $f' + f/r = -\kappa\tau^2\rho_{eff}r$, respectively. Using the values of $\kappa = 8\pi G/c^2\tau^2$ and $\rho_c = 3/8\pi G\tau^2$, we obtain

$$f\left(r\right) = \left(1 - \Omega_m\right)r^2/c^2\tau^2, \qquad (4.13)$$

where $\Omega_m = \rho/\rho_c$.

4.4.2 Pressure

We also obtain for the pressure

$$p = \frac{1 - \Omega_m}{\kappa c \tau^3} = \frac{c}{\tau} \frac{1 - \Omega_m}{8\pi G}, \tag{4.14}$$

$$e^{-\mu} = 1 + f(r) = 1 + \tau c^{-1} \kappa p r^2 = 1 + (1 - \Omega_m) r^2 / c^2 \tau^2. \tag{4.15}$$

4.4.3 Line element

Accordingly, the line element of the Universe is given by

$$ds^2 = \tau^2 dv^2 - \frac{dr^2}{1 + (1 - \Omega_m) r^2 / c^2 \tau^2} - r^2 \left(d\theta^2 + \sin^2 \theta d\phi^2 \right), \tag{4.16}$$

or,

$$ds^2 = \tau^2 dv^2 - \frac{dr^2}{1 + (\kappa \tau / c) p r^2} - r^2 \left(d\theta^2 + \sin^2 \theta d\phi^2 \right). \tag{4.17}$$

This line element is the comparable to the FRW line element in the standard theory.

It will be recalled that the Universe expansion is determined by Eq. (4.4), $dr/dv = \tau e^{-\mu/2}$. The only thing that is left to be determined is the sign of $(1 - \Omega_m)$ or the pressure p. Thus we have

$$\frac{dr}{dv} = \tau \sqrt{1 + \kappa \tau c^{-1} p r^2} = \tau \sqrt{1 + \frac{1 - \Omega_m}{c^2 \tau^2} r^2}. \tag{4.18}$$

4.5 Physical meaning

We now classify the Universe by the values of Ω_m.

4.5.1 $\Omega_m > 1$:

For $\Omega_m > 1$ one obtains

$$r(v) = \frac{c\tau}{\alpha} \sin \alpha \frac{v}{c}, \qquad \alpha = \sqrt{\Omega_m - 1}. \tag{4.19}$$

This is obviously a closed Universe, and presents a decelerating expansion.

4.5.2 $\Omega_m < 1$:

For $\Omega_m < 1$ one obtains

$$r(v) = \frac{c\tau}{\beta} \sinh \beta \frac{v}{c}, \qquad \beta = \sqrt{1 - \Omega_m}. \tag{4.20}$$

This is now an open accelerating Universe.

4.5.3 $\Omega_m = 1$:

For $\Omega_m = 1$ we have, of course, $r = \tau v$.

4.6   The accelerating Universe

From the above one can write the expansion of the Universe in the standard Hubble form $v = H_0 r$ with

$$H_0 = h \left[1 - (1 - \Omega_m) v^2 / 6c^2 \right], \tag{4.21}$$

where $h = \tau^{-1}$. Thus H_0 depends on the distance that is being measured. It is well-known that the farther the distance, the lower the value for H_0 is measured. This is possible only for $\Omega_m < 1$, i.e. when the Universe is accelerating. In that case the pressure is positive.

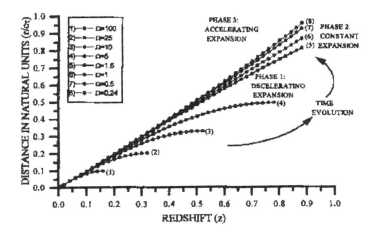

Figure 4.1: Hubble's diagram describing the tri-phase evolution of the Universe according to cosmological general relativity theory (adapted from Behar and Carmeli).

4.6.1   Tri-phase expansion

Figure 4.1 describes the Hubble diagram of the above solutions for the three types of expansion for values of Ω_m from 100 to 0.245. The figure describes the three-phase evolution of the Universe. Curves (1)-(5) represent the stages of *decelerating expansion* according to Eq. (4.19). As the density of matter ρ decreases, the Universe goes over from the lower curves to the upper ones, but it does not have enough time to close up to a Big Crunch. The Universe subsequently goes over to curve (6) with $\Omega_m = 1$, at which time it has a constant expansion for a fraction of a second. This then followed by going to the upper curves (7) and (8) with $\Omega_m < 1$, where the Universe expands with *acceleration* according to Eq. (4.20). Curve no. 8 fits the present situation of the Universe. For curves (1)-(4) in the diagram we use the cutoff when the curves were at their maximum.

4.7   Theory versus experiment

4.7.1   Value of the Big Bang time τ

To find out the numerical value of τ we use the relationship between $h = \tau^{-1}$ and H_0 given by Eq. (4.21)(CR denote values according to Cosmological Relativity):

$$H_0 = h\left[1 - \left(1 - \Omega_m^{CR}\right) z^2/6\right],\qquad(4.22)$$

where $z = v/c$ is the redshift and $\Omega_m^{CR} = \rho_m/\rho_c$ with $\rho_c = 3h^2/8\pi G$. (Notice that our ρ_c is different from the standard ρ_c defined with H_0 as $3H_0^2/8\pi G$.) The redshift parameter z determines the distance at which H_0 is measured. We choose $z = 1$ and take for $\Omega_m^{CR} = 0.245$, its value at the present time (see Table 4.1) (corresponds to 0.32 in the standard theory), Eq. (4.22) then gives $H_0 = 0.874h$.

Based on curve fitting equations to all the data in the plot, Hartnett obtained (see Fig. 4.2) $h = 72.47 \pm 1.95$km/s-Mpc, from Tully-Fisher (TF) (the solid line), and $h = 72.17 \pm 0.84$km/s-Mpc, from SNe type Ia (the broken line) measurements. The error bars here are statistical fit standard errors, but the rms errors on the published data are ± 1.64km/s-Mpc for SNe type Ia determined and ± 13.24 for TF determined. We choose

$$h = 72.17 \pm 0.84 \pm 1.64\text{km/s-Mpc}.\qquad(4.23)$$

Therefore

$$\tau = (4.28 \pm 0.15) \times 10^{17}s = 13.56 \pm 0.48\text{Gyr}.\qquad(4.24)$$

This result fits the recent obtained measurements by NASA's WMAP, according to which initial cosmic inflation happened 13.7 ± 0.2Gyr ago.

Table 4.1: The Cosmic Times with respect to the Big Bang, the Cosmic Temperature and the Cosmic Pressure for each of the Curves in Fig. 4.1.

Curve No*.	Ω_m	Time in Units of τ	Time (Gyr)	Temperature (K)	Pressure (g/cm^2)
		DECELERATING EXPANSION			
1	100	3.1×10^{-6}	4.23×10^{-5}	1096	-4.142
2	25	9.8×10^{-5}	1.32×10^{-3}	195.0	-1.004
3	10	3.0×10^{-4}	4.07×10^{-3}	111.5	-0.377
4	5	1.2×10^{-3}	1.63×10^{-2}	58.20	-0.167
5	1.5	1.3×10^{-2}	1.76×10^{-1}	16.43	-0.021
		CONSTANT EXPANSION			
6	1	3.0×10^{-2}	4.07×10^{-1}	11.15	0
		ACCELERATING EXPANSION			
7	0.5	1.3×10^{-1}	1.76	5.538	$+0.021$
8	0.245	1.0	13.56	2.730	$+0.032$

*The calculations are made using the cosmological transformation, Eq. (2.10), that relates physical quantities at different cosmic times when gravity is extremely weak.

For example, we denote the temperature by θ, and the temperature at the present time by θ_0, we then have

$$\theta = \frac{\theta_0}{\sqrt{1 - \dfrac{t^2}{\tau^2}}} = \frac{\theta_0}{\sqrt{1 - \dfrac{(\tau - T)^2}{\tau^2}}} = \frac{2.73K}{\sqrt{\dfrac{2\tau T - T^2}{\tau^2}}}$$

$$= \frac{2.73K}{\sqrt{\dfrac{T}{\tau}\left(2 - \dfrac{T}{\tau}\right)}},$$

where T is the time with respect to the Big Bang time.

The formula for the pressure is given by Eq. (4.14), $p = c(1 - \Omega_m)/8\pi G\tau$. Using $c = 3 \times 10^{10} cm/s$, $\tau = 4.28 \times 10^{17} s$ and

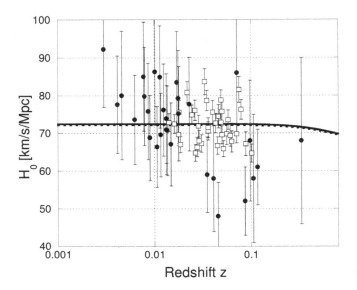

Figure 4.2: Hartnett's diagram for determining the Big Bang time $\tau (= 13.56 \pm 0.48\text{Gyr})$.

$G = 6.67 \times 10^{-8} cm^3/gs^2$, we obtain

$$p = 4.184 \times 10^{-2} \left(1 - \Omega_m\right) g/cm^2.$$

4.7.2   Value of Ω_Λ^{CR}

What is left is to find the value of Ω_Λ^{CR}. We have $\Omega_\Lambda^{CR} = \rho_c^{ST}/\rho_c$, where $\rho_c^{ST} = 3H_0^2/8\pi G$ and $\rho_c = 3h^2/8\pi G$. Thus $\Omega_\Lambda^{CR} = (H_0/h)^2 = 0.874^2$, or

$$\Omega_\Lambda^{CR} = 0.764. \tag{4.25}$$

As is seen from the above equations one has

$$\Omega_T = \Omega_m^{CR} + \Omega_\Lambda^{CR} = 0.245 + 0.764 = 1.009 \approx 1, \tag{4.26}$$

which means the Universe is Euclidean.

Table 4.2: Cosmological parameters in cosmological general relativity and in standard theory.

	Cosmological Relativity	Standard Theory
Theory type	Spacevelocity	Spacetime
Expansion type	Tri-phase: decelerating, constant, accelerating	One phase
Inflation	Follows from theory	Assumed
Present expansion	Accelerating (predicted)	One of three possibilities
Pressure	Positive	Negative
Cosmological constant	None	Depends
$\Omega_T = \Omega_m + \Omega_\Lambda$	1.009	Depends
Constant-expansion occurs at	8.5Gyr ago (Gravity included)	No prediction
Constant-expansion duration	Fraction of second	Not known
Temperature at constant expansion	146K (Gravity included)	No prediction

Our results confirm those of the supernovae experiments and indicate on the existence of the dark energy as has recently received confirmation from the Boomerang cosmic microwave background experiment, which showed that the Universe is Euclidean.

Table 4.2 gives some of the cosmological parameters obtained in this theory and compares them to those obtained in the standard theory.

4.8   Comparison with general relativity

One has to add the time dimension and the result is a five-dimensional theory of space-time-velocity. One can show that all the classical experiments predicted by general relativity are also predicted and satisfied by CGR (see Chapter 5).

4.9   Recent developments on dark matter

Using the theory presented here, John Hartnett has recently shown that there is no need for the assumption of dark matter in spiral galaxies.

References

S. Behar and M. Carmeli, "Cosmological Relativity: A New Theory of Cosmology," *Intern. J. Theor. Phys.* **39**, 1375 (2000), astro-ph/0008352.

P. de Bernardis *et al.*, *Nature* **404**, 955 (2000), astro-ph/0004404

M. Carmeli, *Classical Fields: General Relativity and Gauge Theory*, John Wiley-Interscience, New York, 1982. Reprinted by World Scientific, 2001.

M. Carmeli, "Cosmological General Relativity," *Communications in Theoretical Physics* **5**, 159 (1996).

M. Carmeli, *Cosmological Special Relativity: The Large-Scale Structure of Space, Time and Velocity*, Second Edition, World Scientific, Singapore, 2002.

A.V. Filippenko and A.G. Riess, p.227 in: *Particle Physics and Cosmology: Second Tropical Workshop*, J.F. Nieves, Editor, AIP, New York (2000)

J. Hartnett, "Carmeli's accelerating Universe is spatially flat without dark matter", *Intern. J. Theor. Phys.* **44** (4), 485 - 492 (2005), gr-qc/0407083

J.G. Hartnett, "The Carmeli metric correctly describes spiral galaxy rotation curves" *Intern. J. Theor. Phys.* **44** (3), 359 - 372 (2005), gr-qc/0407082

J.G. Hartnett, "Carmeli's cosmology: The Universe is spatially flat without dark matter", invited talk given at the international conference "Frontiers of Fundamental Physics 6", held in Udine, Italy, September 26-29, 2004

J.G. Hartnett, Personal letters to the author 1 and 5 March 2006.

P.J.E. Peebles, Status of the big bang cosmology, in: *Texas/ Pascos 92: Relativistic Astrophysics and Particle Cosmology*, C.W. Akerlof and M.A. Srednicki, Editors, p. 84, New York Academy of Sciences, New York (1993)

Physics Today, "Search and Discovery", May 2006, pp. 16 - 18.

CHAPTER 5

COSMOLOGICAL GENERAL RELATIVITY IN FIVE DIMENSIONS

5.1 Introduction

In this chapter we present the five-dimensional cosmological general relativity of space, time and velocity. The added extra dimension of velocity to the usual four-dimensional spacetime will be evident in the sequel. Important basic issues that we face in five dimensions are also discussed.

5.1.1 Five-dimensional manifold of space, time and velocity

If we add the time to the cosmological flat spacevelocity line element, we obtain (see Sections 2.11 and 2.12)

$$ds^2 = c^2dt^2 - (dx^2 + dy^2 + dz^2) + \tau^2 dv^2. \qquad (5.1)$$

Accordingly, we have a five-dimensional manifold of time, space and velocity. The above line element provides a group of transformations O(2,3). At v=const it yields the Minkowskian line element; at t=const it gives the cosmological line element; and at a fixed space point, $dx = dy = dz = 0$, it leads to a new two-dimensional line element

$$ds^2 = c^2 dt^2 + \tau^2 dv^2. \tag{5.2}$$

The groups associated with the above line elements are, of course, O(1,3), O(3,1) and O(2), respectively. They are the Lorentz group, the cosmological group and a two-dimensional Euclidean group, respectively.

In the next section we discuss some properties of the Universe with gravitation in five dimensions. That includes the Bianchi identities, the gravitational field equations, the velocity as an independent coordinate and the energy density in cosmology. We then find the equations of motion of the expanding Universe and show that the Universe is accelerating. Afterward we discuss the important problem of halo dark matter around galaxies by finding the equations of motion of a star moving around a spherically-symmetric galaxy. The equations obtained are *not* Newtonian and instead the Tully-Fisher formula is obtained from our theory. We then show that the Universe is infinite and open, now by applying redshift analysis, using a new formula that is derived here. Finally, the concluding remarks and some mathematical conventions are presented.

5.2   Universe with gravitation

The Universe is, of course, not flat but filled up with gravity. When gravitation is invoked, the above spaces become curved Riemanian with the line element

$$ds^2 = g_{\mu\nu} dx^\mu dx^\nu,$$

where μ, ν take the values 0, 1, 2, 3, 4. The coordinates are: $x^0 = ct$, x^1, x^2, x^3 are spatial coordinates and $x^4 = \tau v$ (the role of the velocity as an independent coordinate will be discussed in the sequel). The signature is $(+ - - - +)$. The metric tensor $g_{\mu\nu}$

is symmetric and thus we have fifteen independent components. They will be a solution of the Einstein field equations in five dimensions. A discussion on the generalization of the Einstein field equations from four to five dimensions will also be given.

5.2.1 The Bianchi identities

The restricted Bianchi identities are given by

$$\left(R^\nu_\mu - \frac{1}{2}\delta^\nu_\mu R \right)_{;\nu} = 0, \tag{5.3}$$

where $\mu,\nu=0,\ldots,4$. They are valid in five dimensions just as they are in four dimensions. In Eq. (5.3) R^ν_μ and R are the Ricci tensor and scalar, respectively, and a semicolon denotes covariant differentiation. As a consequence we now have five coordinate conditions that permit us to determine five coordinates. For example, one can choose $g_{00} = 1$, $g_{0k} = 0$, $g_{44} = 1$, where $k=1, 2, 3$. These are the co-moving coordinates in five dimensions that keep the clocks and the velocity-measuring instruments synchronized. We will not use these coordinates in this chapter.

5.2.2 The gravitational field equations

In four dimensions these are the Einstein field equations:

$$R_{\mu\nu} - \frac{1}{2}g_{\mu\nu}R = \kappa T_{\mu\nu}, \tag{5.4}$$

or equivalently

$$R_{\mu\nu} = \kappa \left(T_{\mu\nu} - \frac{1}{2}g_{\mu\nu}T \right), \tag{5.5}$$

where $T = g_{\alpha\beta}T^{\alpha\beta}$, and we have $R = -\kappa T$. In five dimensions if one chooses Eq. (5.4) as the field equations then Eq. (5.5) is *not* valid (the factor $\frac{1}{2}$ will have to be replaced by $\frac{1}{3}$, and $R = -\frac{2}{3}\kappa T$), and thus there is no symmetry between R and $-\kappa T$.

5.2.3 Velocity as an independent coordinate

First we have to iterate what do we mean by coordinates in general and how one measures them. The time coordinate is measured by clocks as was emphasized by Einstein repeatedly. So are the spatial coordinates: they are measured by meters, as was originally done in special relativity theory by Einstein, or by use of Bondi's more modern version of k-calculus.

But how about the velocity as an independent coordinate? One might incline to think that if we know the spatial coordinates then the velocities are just the time derivative of the coordinates and they are not independent coordinates. This is, indeed, the situation for a dynamical system when the coordinates are given as functions of the time. But in general the situation is different, especially in cosmology. Take, for instance, the Hubble law $v = H_0 x$. Obviously v and x are independent parameters and v is not the time derivative of x. Basically one can measure v by instruments like those used by traffic police.

5.2.4 Effective mass density in cosmology

To finish this section we discuss the important concept of the energy density in cosmology. We use the Einstein field equations, in which the right-hand side includes the energy-momentum tensor. For fields other than gravitation, like the electromagnetic field, this is a straightforward expression that comes out as a generalization to curved spacetime of the same tensor appearing in special-relativistic electrodynamics. However, when dealing with matter one should construct the energy-momentum tensor according to the physical situation (see, for example, Fock). Often a special expression for the mass density ρ is taken for the right-hand side of Einstein's equations, which sometimes is expressed as a δ-function.

In cosmology we also have the situation where the mass density is put on the right-hand side of the Einstein field equations.

There is also the critical mass density $\rho_c = 3/8\pi G\tau^2$, the value of which is about 10^{-29} g/cm^3, just a few hydrogen atoms per cubic meter throughout the cosmos. If the universe average mass density ρ is equal to ρ_c then the three spatial geometry of the four-dimensional cosmological space is Euclidean. A deviation from this Euclidean geometry necessitates an increase or decrease from ρ_c. That is to say that

$$\rho_{eff} = \rho - \rho_c \tag{5.6}$$

is the active or the effective mass density that causes the three geometry not to be Euclidean. Accordingly, one should use ρ_{eff} in the right-hand side of the Einstein field equations. Indeed, we will use such a convention throughout this chapter. The subtraction of ρ_c from ρ in not significant for celestial bodies and makes no difference.

5.3 The accelerating Universe

5.3.1 Preliminaries

In the last two sections we gave arguments to the fact that the Universe should be presented in five dimensions, even though the standard cosmological theory is obtained from Einstein's four-dimensional general relativity theory. The situation here is similar to that prevailed before the advent of ordinary special relativity. At that time the equations of electrodynamics, written in three dimensions, were well known to predict that the speed of light was constant. But that was not the end of the road. The abandon of the concept of absolute space along with the constancy of the speed of light led to the four-dimensional notion. In cosmology now, we have to give up the notion of absolute cosmic time. Then this, with the constancy of the Big Bang time τ, leads us to a five-dimensional presentation of cosmology.

We recall that the field equations are those of Einstein in five dimensions,

$$R_\mu^\nu - \frac{1}{2}\delta_\mu^\nu R = \kappa T_\mu^\nu, \tag{5.4}$$

where Greek letters $\alpha, \beta, \cdots, \mu, \nu, \cdots = 0, 1, 2, 3, 4$. The coordinates are $x^0 = ct$, x^1, x^2 and x^3 are space-like coordinates, $r^2 = (x^1)^2 + (x^2)^2 + (x^3)^2$, and $x^4 = \tau v$. The metric used is given by (see Appendix A)

$$g_{\mu\nu} = \begin{pmatrix} 1+\phi & 0 & 0 & 0 & 0 \\ 0 & -1 & 0 & 0 & 0 \\ 0 & 0 & -1 & 0 & 0 \\ 0 & 0 & 0 & -1 & 0 \\ 0 & 0 & 0 & 0 & 1+\psi \end{pmatrix}, \tag{5.7}$$

We will keep only linear terms. The nonvanishing Christoffel symbols are given by (see Appendix A)

$$\Gamma_{0\lambda}^0 = \frac{1}{2}\phi_{,\lambda}, \ \Gamma_{44}^0 = -\frac{1}{2}\psi_{,0}, \ \Gamma_{00}^n = \frac{1}{2}\phi_{,n},$$

$$\Gamma_{44}^n = \frac{1}{2}\psi_{,n}, \ \Gamma_{00}^4 = -\frac{1}{2}\phi_{,4}, \ \Gamma_{4\lambda}^4 = \frac{1}{2}\psi_{,\lambda},$$

where $n = 1, 2, 3$ and a comma denotes partial differentiation. The components of the Ricci tensor and the Ricci scalar are given by (Appendix A)

$$R_0^0 = \frac{1}{2}\left(\nabla^2\phi - \phi_{,44} - \psi_{,00}\right), \tag{5.8a}$$

$$R_0^n = \frac{1}{2}\psi_{,0n}, \quad R_n^0 = -\frac{1}{2}\psi_{,0n}, \quad R_0^4 = R_4^0 = 0, \tag{5.8b}$$

$$R_m^n = \frac{1}{2}\left(\phi_{,mn} + \psi_{,mn}\right), \tag{5.8c}$$

$$R_n^4 = -\frac{1}{2}\phi_{,n4}, \quad R_4^n = \frac{1}{2}\phi_{,n4}. \tag{5.8d}$$

$$R_4^4 = \frac{1}{2}\left(\nabla^2\psi - \phi_{,44} - \psi_{,00}\right), \tag{5.8e}$$

$$R = \nabla^2\phi + \nabla^2\psi - \phi_{,44} - \psi_{,00}. \tag{5.9}$$

In the above equations ∇^2 is the ordinary three-dimensional Laplace operator.

5.3.2 Expanding Universe

The line element in five dimensions is given by

$$ds^2 = (1 + \phi)dt^2 - dr^2 + (1 + \psi)dv^2, \tag{5.10}$$

where $dr^2 = (dx^1)^2 + (dx^2)^2 + (dx^3)^2$, and where c and τ were taken, for brevity, as equal to 1. For an expanding Universe one has $ds = 0$. The line element (5.10) represents a spherically symmetric Universe.

The expansion of the Universe (the Hubble expansion) is recorded at a definite instant of time and thus $dt = 0$. Accordingly, taking into account $d\theta = d\phi = 0$, Eq. (5.10) gives the following equation for the expansion of the Universe at a certain moment,

$$-dr^2 + (1 + \psi)dv^2 = 0, \tag{5.11}$$

and thus

$$\left(\frac{dr}{dv}\right)^2 = 1 + \psi. \tag{5.12}$$

To find ψ we solve the Einstein field equation (noting that $T_0^0 = g_{0\alpha}T^{\alpha 0} \approx T^{00} = \rho(dx^0/ds)^2 \approx c^2\rho$, or $T_0^0 \approx \rho$ in units with $c = 1$):

$$R_0^0 - \frac{1}{2}\delta_0^0 R = 8\pi G\rho_{eff} = 8\pi G\left(\rho - \rho_c\right), \tag{5.13}$$

where $\rho_c = 3/8\pi G\tau^2$.

A simple calculation then yields

$$\nabla^2\psi = 6(1 - \Omega), \tag{5.14}$$

where $\Omega = \rho/\rho_c$.

The solution of the field equation (5.14) is given by

$$\psi = (1 - \Omega)r^2 + \psi_0, \tag{5.15}$$

where the first part on the right-hand side is a solution for the non-homogeneous Eq. (5.14), and ψ_0 represents a solution to its homogeneous part, i.e. $\nabla^2 \psi_0 = 0$. A solution for ψ_0 can be obtained as an infinite series in powers of r. The only term that is left is of the form $\psi_0 = -K_2/r$, where K_2 is a constant whose value can easily be shown to be the Schwartzschild radius, $K_2 = 2GM$. We therefore have

$$\psi = (1 - \Omega)r^2 - 2GM/r. \tag{5.16}$$

The Universe expansion is therefore given by

$$\left(\frac{dr}{dv}\right)^2 = 1 + (1 - \Omega)\, r^2 - \frac{2GM}{r}. \tag{5.17}$$

For large r the last term on the right-hand side can be neglected, and therefore

$$\left(\frac{dr}{dv}\right)^2 = 1 + (1 - \Omega)r^2, \tag{5.18}$$

or

$$\frac{dr}{dv} = \left[1 + (1 - \Omega)\, r^2\right]^{1/2}. \tag{5.19}$$

Inserting now the constants c and τ we finally obtain for the expansion of the Universe

$$\frac{dr}{dv} = \tau \left[1 + (1 - \Omega)\, r^2/c^2\tau^2\right]^{1/2}. \tag{5.20}$$

The second term in the square bracket of Eq. (5.20) represents the deviation from constant expansion due to gravity. For without this term, Eq. (5.20) reduces to $dr/dv = \tau$, thus $r = \tau v + const$. The constant can be taken zero if one assumes, as usual, that at $r = 0$ the velocity should also vanish. Accordingly we have $r = \tau v$ or $v = \tau^{-1}r$. Hence when $\Omega = 1$, namely when $\rho = \rho_c$, we have a constant expansion.

5.3.3 Decelerating, constant and accelerating expansions

The equation of motion (5.20) can be integrated exactly (see Appendix B). We have three cases:

Case 1:

For the $\Omega > 1$ case

$$r(v) = (c\tau/\alpha) \sin(\alpha v/c); \quad \alpha = (\Omega - 1)^{1/2}. \quad (5.21a)$$

This is obviously a decelerating expansion.

Case 2:

For $\Omega < 1$,

$$r(v) = (c\tau/\beta) \sinh(\beta v/c); \quad \beta = (1 - \Omega)^{1/2}. \quad (5.21b)$$

This is now an accelerating expansion.

Case 3:

For $\Omega = 1$ we have, from Eq. (5.20),

$$d^2r/dv^2 = 0, \quad (5.21c)$$

whose solution is, of course,

$$r(v) = \tau v, \quad (5.22)$$

and this is a constant expansion. It will be noted that the last solution can also be obtained directly from the previous two cases for $\Omega > 1$ and $\Omega < 1$ by going to the limit $v \to 0$, using L'Hospital's lemma, showing that our solutions are consistent.

It has been shown that the constant expansion is just a transition stage between the decelerating and the accelerating expansions as the Universe evolves toward its present situation. This occurred at 8.5 Gyr ago at a time the cosmic radiation temperature was 143K.

5.3.4 Accelerating Universe

In order to decide which of the three cases is the appropriate one at the present time, it will be convenient to write the solutions in the ordinary Hubble law form $v = H_0 r$. Expanding Eqs. (5.21a,b) and keeping the appropriate terms then yields

$$r = \tau v \left(1 - \alpha^2 v^2 / 6c^2\right), \tag{5.23}$$

$$r = \tau v \left(1 + \beta^2 v^2 / 6c^2\right), \tag{5.24}$$

for the $\Omega > 1$ and $\Omega < 1$ cases, respectively. Using now the expressions for α and β, then both of the last equations can be reduced into the single equation

$$r = \tau v \left[1 + (1 - \Omega) v^2 / 6c^2\right]. \tag{5.25}$$

Inverting now this equation by writing it in the form $v = H_0 r$, we obtain in the lowest approximation for H_0,

$$H_0 = h \left[1 - (1 - \Omega) v^2 / 6c^2\right], \tag{5.26}$$

where $h = 1/\tau$. Using $v \approx r/\tau$, or $z \approx v/c$, we also obtain

$$H_0 = h \left[1 - (1 - \Omega) r^2 / 6c^2 \tau^2\right] = h \left[1 - (1 - \Omega) z^2 / 6\right]. \tag{5.27}$$

The above equations show that H_0 depends on the distance, or equivalently, on the redshift. Consequently H_0 has meaning only in the limits $r \to 0$ and $z \to 0$, namely when measured *locally*, in which case it becomes the constant h. This is similar to the situation with respect to the speed of light when measured globally in the presence of gravitational field as the ratio between distance and time, the result usually depends on these parameters. Only in the limit one obtains the constant speed of light in vacuum ($c \approx 3 \times 10^{10}$cm/s).

As is seen from the above discussion, H_0 is intimately related to the sign of the factor $(1 - \Omega)$. If measurements of H_0 indicate that it increases with the redshift parameter z then the sign of $(1 - \Omega)$ is negative, namely $\Omega > 1$. If, however, H_0 decreases when z increases then the sign of $(1 - \Omega)$ is positive, i.e. $\Omega < 1$. The possibility of H_0 not to depend on the redshift parameter indicates that $\Omega = 1$. In recent years different measurements were obtained for H_0, with the so-called "short" and "long" distance scales, in which higher values of H_0 were obtained for the short distances and the lower values for H_0 corresponded to the long distances. Indications are that the longer the distance of measurement, the smaller the value of H_0. If one takes these experimental results seriously, then that is possible only for the case in which $\Omega < 1$, namely when the Universe is at an accelerating expansion phase, and the Universe is thus open. We will see in Section 5.6 that the same result is obtained via a new cosmological redshift formula.

5.4 The Tully-Fisher formula: Nonexistence of halo dark matter

In this section we derive the equations of motion of a star moving around a spherically symmetric galaxy and show that the Tully-Fisher formula is obtained from the five-dimensional cosmological general relativity theory. The calculation is lengthy but it is straightforward. The equations of motion will first be of general nature and only afterward specialized to the motion of a star around the field of a galaxy. The equations obtained are *not* Newtonian. The Tully-Fisher formula was obtained previously using two representations of Einstein's general relativity: the standard spacetime theory and a spacevelocity version of it. However, the present derivation is a straightforward result from the unification of space, time and velocity.

Our notation in this section is as follows: $\alpha, \beta, \gamma, \ldots = 0, \cdots$, 4; $a, b, c, d, \cdots = 0, \cdots, 3$; $p, q, r, s, \cdots = 1, \cdots, 4$; and k, l, m, $n, \cdots = 1, 2, 3$. The coordinates are: $x^0 = ct$ (timelike), $x^k = x^1, x^2, x^3$ (spacelike), and $x^4 = \tau v$ (velocitylike).

5.4.1   The Geodesic Equation

As usual the equations of motion are obtained in general relativity theory from the covariant conservation law of the energy-momentum tensor (which is a consequence of the restricted Bianchi identities), and the result, as is well known, is the geodesic equation that describes the motion of a spherically symmetric test particle. In our five-dimensional cosmological theory we have five equations of motion. They are given by

$$\frac{d^2 x^\mu}{ds^2} + \Gamma^\mu_{\alpha\beta} \frac{dx^\alpha}{ds} \frac{dx^\beta}{ds} = 0. \tag{5.28}$$

We now change the independent parameter s into an arbitrary new parameter σ, then the geodesic equation becomes

$$\frac{d^2 x^\mu}{d\sigma^2} + \Gamma^\mu_{\alpha\beta} \frac{dx^\alpha}{d\sigma} \frac{dx^\beta}{d\sigma} = -\frac{d^2\sigma/ds^2}{(d\sigma/ds)^2} \frac{dx^\mu}{d\sigma}. \tag{5.29}$$

The parameter σ will be taken once as $\sigma = x^0$ (the time coordinate) and then $\sigma = x^4$ (the velocity coordinate). We obtain, for the first case,

$$\frac{d^2 x^p}{(dx^0)^2} + \left(\Gamma^p_{\alpha\beta} - \Gamma^0_{\alpha\beta} \frac{dx^p}{dx^0} \right) \frac{dx^\alpha}{dx^0} \frac{dx^\beta}{dx^0} = 0, \tag{5.30}$$

where $p = 1, 2, 3, 4$.

In exactly the same way we parametrize the geodesic equation now with respect to the velocity by choosing the parameter $\sigma = \tau v$. The result is

$$\frac{d^2x^a}{(dx^4)^2} + \left(\Gamma^a_{\alpha\beta} - \Gamma^4_{\alpha\beta}\frac{dx^a}{dx^4}\right)\frac{dx^\alpha}{dx^4}\frac{dx^\beta}{dx^4} = 0, \qquad (5.31)$$

where $a = 0, 1, 2, 3$.

The equation of motion (5.30) will be expanded in terms of the parameter v/c, assuming $v \ll c$, whereas Eq. (5.31) will be expanded with t/τ, where t is a characteristic cosmic time, and $t \ll \tau$. We then can use the Einstein-Infeld-Hoffmann (EIH) method that is well known in general relativity in obtaining the equations of motion.

We obtain

$$\frac{d^2x^p}{dt^2} + \Gamma^p_{\alpha\beta}\frac{dx^\alpha}{dt}\frac{dx^\beta}{dt} = 0, \qquad (5.32)$$

$$\frac{d^2x^a}{dv^2} + \Gamma^a_{\alpha\beta}\frac{dx^\alpha}{dv}\frac{dx^\beta}{dv} = 0. \qquad (5.33)$$

To find the lowest approximation of Eq. (5.32), since $dx^0/dt \gg dx^q/dt$, all terms with indices that are not zero-zero can be neglected. Consequently, Eq. (5.32) is reduced to the form

$$\frac{d^2x^p}{dt^2} \approx -\Gamma^p_{00}, \qquad (5.34)$$

in the lowest approximation.

5.4.2 Equations of motion

Accordingly Γ^p_{00} acts like a Newtonian force per mass unit. In terms of the metric tensor we therefore obtain, since $\Gamma^p_{00} = -\frac{1}{2}\eta^{pq}\phi_{,q}$

$$\frac{d^2x^p}{dt^2} \approx -\frac{1}{2}\eta^{pq}\frac{\partial\phi}{\partial q}, \qquad (5.35)$$

where $\phi = g_{00} - 1$. We now decompose this equation into a spatial ($p = 1, 2, 3$) and a velocity ($p = 4$) parts, getting

$$\frac{d^2 x^k}{dt^2} = -\frac{1}{2}\frac{\partial \phi}{\partial x^k}, \tag{5.36a}$$

$$\frac{d^2 v}{dt^2} = 0. \tag{5.36b}$$

Using exactly the same method, Eq. (5.33) yields

$$\frac{d^2 x^k}{dv^2} = -\frac{1}{2}\frac{\partial \psi}{\partial x^k}, \tag{5.37a}$$

$$\frac{d^2 t}{dv^2} = 0, \tag{5.37b}$$

where $\psi = g_{44} - 1$. In the above equations $k = 1, 2, 3$. Equation (5.36a) is exactly the law of motion with the function ϕ being twice the Newtonian potential. The other three equations Eq. (5.36b) and Eqs. (5.37a,b) are not Newtonian and are obtained only in the present theory. It remains to find out the functions ϕ and ψ.

The functions ϕ and ψ

To find out the function ϕ we solve the Einstein field equation (noting that $T_4^4 = g_{4\alpha}T^{\alpha 4} \approx T^{44} = \rho(dx^4/ds)^2 \approx \tau^2\rho$, and thus $T_4^4 \approx \rho$ in units in which $\tau = 1$):

$$R_4^4 - \frac{1}{2}\delta_4^4 R = 8\pi G\rho_{eff} = 8\pi G(\rho - \rho_c). \tag{5.38}$$

A straightforward calculation then gives

$$\nabla^2\phi = 6\left(1 - \Omega\right), \tag{5.39}$$

whose solution is given by

$$\phi = (1 - \Omega)\, r^2 + \phi_0, \tag{5.40}$$

where ϕ_0 is a solution of the homogeneous equation $\nabla^2 \phi_0 = 0$. One then easily finds that $\phi_0 = -K_1/r$, where $K_1 = 2GM$. In the same way the function ψ can be found (see Section 5.3),

$$\psi = (1 - \Omega)\, r^2 + \psi_0, \tag{5.41}$$

with $\nabla^2 \psi_0 = 0$, $\psi_0 = -K_2/r$ and $K_2 = 2GM$. (When units are inserted then $K_1 = 2GM/c^2$ and $K_2 = 2GM\tau^2/c^2$.) For the purpose of obtaining equations of motion one can neglect the terms $(1 - \Omega)r^2$, actually $(1 - \Omega)r^2/c^2\tau^2$, in the solutions for ϕ and ψ. One then obtains

$$g_{00} \approx 1 - 2GM/c^2 r, \quad g_{44} \approx 1 - 2GM\tau^2/c^2 r. \tag{5.42}$$

The equations of motion, consequently, have the forms, when inserting the constants c and τ,

$$\frac{d^2 x^k}{dt^2} = GM \left(\frac{1}{r}\right)_{,k}, \tag{5.43a}$$

$$\frac{d^2 x^k}{dv^2} = kM \left(\frac{1}{r}\right)_{,k}, \tag{5.43b}$$

where $k = G\tau^2/c^2$. It remains to integrate equations (5.36b) and (5.37b). One finds that $v = a_0 t$, where a_0 is a constant which can be taken as equal to $a_0 = c/\tau \approx cH_0$. Accordingly, we see that the particle experiences an acceleration $a_0 = c/\tau \approx cH_0$.

Equation (5.43a) is Newtonian but (5.43b) is not. The integration of the latter is identical to that familiar in classical Newtonian mechanics, but there is an essential difference which should be emphasized. In Newtonian equations of motion one deals with a path of motion in the 3-space. In our theory we do not have that situation. Rather, the paths here indicate locations of particles in the sense of the Hubble distribution, which now takes a different physical meaning. With that in mind we proceed as follows.

Equation (5.43b) yields the first integral

$$\left(\frac{ds}{dv}\right)^2 = \frac{kM}{r},\tag{5.44a}$$

where v is the velocity of the particles, in analogy to the Newtonian

$$\left(\frac{ds}{dt}\right)^2 = \frac{GM}{r}.\tag{5.44b}$$

In these equations s is the length parameter along the path of the accumulation of the particles.

Comparing Eqs. (5.44a) and (5.44b), we obtain

$$\frac{ds}{dv} = \frac{\tau}{c}\frac{ds}{dt}.\tag{5.45}$$

Thus

$$\frac{dv}{dt} = \frac{c}{\tau}.\tag{5.46}$$

Accordingly, as we have mentioned before, the particle experiences an acceleration $a_0 = c/\tau \approx cH_0$.

5.4.3   The Tully-Fisher law

The motion of a particle in a central field is best described in terms of an "effective potential", V_{eff}. In Newtonian mechanics this is given by

$$V_{eff} = -\frac{GM}{r} + \frac{L^2}{2r^2},\tag{5.47}$$

where L is the angular momentum per mass unit. In our case the effective potential is

$$V_{eff}(r) = -\frac{GM}{r} + \frac{L^2}{2r^2} + a_0 r.\tag{5.48}$$

The circular motion is obtained at the minimal value of (5.48), i.e.,

$$\frac{dV_{eff}}{dr} = \frac{GM}{r^2} - \frac{L^2}{r^3} + a_0 = 0, \tag{5.49}$$

with $L = v_c r$, and v_c is the circular velocity. This gives

$$v_c^2 = \frac{GM}{r} + a_0 r. \tag{5.50}$$

Thus

$$v_c^4 = \left(\frac{GM}{r}\right)^2 + 2GMa_0 + a_0^2 r^2, \tag{5.51}$$

where $a_0 = c/\tau \approx cH_0$.

The first term on the right-hand side of this equation is purely Newtonian, and cannot be avoided by any reasonable theory. The second one is the Tully-Fisher term. It is well known that astronomical observations show that for disk galaxies the fourth power of the circular velocity of stars moving around the core of the galaxy, v_c^4, is proportional to the total luminosity L of the galaxy to an accuracy of more than two orders of magnitude in L, namely $v_c^4 \propto L$. Since L is proportional to the mass M of the galaxy, one obtains $v_c^4 \propto M$. This is the Tully-Fisher law. There is no dependence on the distance of the star from the center of the galaxy as Newton's law $v_c^2 = GM/r$ requires for circular motion. In order to rectify this deviation from Newton's laws, astronomers assume the existence of halos around the galaxy which are filled with dark matter and arranged in such a way so as to satisfy the Tully-Fisher law for each particular situation.

In conclusion it appears that there is no necessity for the assumption of the existence of halo dark matter around galaxies. Rather, the result can be described in terms of the properties of spacetimevelocity.

5.5   Cosmological redshift analysis

5.5.1   The redshift formula

In this section we derive a general formula for the redshift in which the term $(1-\Omega)$ appears explicitly. Since there are enough data of measurements of redshifts, this allows one to determine what is the sign of $(1 - \Omega)$, positive, zero or negative. Our conclusion is that $(1 - \Omega)$ cannot be negative or zero. This means that the Universe is infinite, and expands forever, a result favored by some cosmologists. To this end we proceed as follows.

Having the metric tensor from Section 5.4 we may now find the redshift of light emitted in the cosmos. As usual, at two points 1 and 2 we have for the wave lengths and frequencies:

$$\frac{\lambda_2}{\lambda_1} = \frac{\nu_1}{\nu_2} = \frac{ds\,(2)}{ds\,(1)} = \sqrt{\frac{g_{00}\,(2)}{g_{00}\,(1)}}. \tag{5.52}$$

Using now the solution for $g_{00} = 1 + \phi$, with ϕ given by Eq. (5.40), in Eq. (5.52), we obtain

$$\frac{\lambda_2}{\lambda_1} = \sqrt{\frac{1 + r_2^2/a^2 - R_s/r_2}{1 + r_1^2/a^2 - R_s/r_1}}. \tag{5.53}$$

In Eq. (5.53) $R_s = 2GM/c^2$ and $a^2 = c^2\tau^2/(1 - \Omega)$.

For a sun-like body with radius R located at the coordinates origin, and an observer at a distance r from the center of the body, we then have $r_2 = r$ and $r_1 = R$, thus

$$\frac{\lambda_2}{\lambda_1} = \sqrt{\frac{1 + r^2/a^2 - R_s/r}{1 + R^2/a^2 - R_s/R}}. \tag{5.54}$$

5.5.2 Particular cases

Since $R \ll r$ and $R_s < R$ is usually the case we can write, to a good approximation,

$$\frac{\lambda_2}{\lambda_1} = \sqrt{\frac{1 + r^2/a^2}{1 - R_s/R}}. \tag{5.55}$$

The term r^2/a^2 in Eq. (5.55) is a pure cosmological one, whereas R_s/R is the standard general relativistic term. For $R \gg R_s$ we then have

$$\frac{\lambda_2}{\lambda_1} = \sqrt{1 + \frac{r^2}{a^2}} = \sqrt{1 + \frac{(1 - \Omega)\, r^2}{c^2 \tau^2}} \tag{5.56}$$

for the pure cosmological contribution to the redshift. If, furthermore, $r \ll a$ we then have

$$\frac{\lambda_2}{\lambda_1} = 1 + \frac{r^2}{2a^2} = 1 + \frac{(1 - \Omega)\, r^2}{2c^2 \tau^2} \tag{5.57}$$

to the lowest approximation in r^2/a^2, and thus

$$z = \frac{\lambda_2}{\lambda_1} - 1 = \frac{r^2}{2a^2} = \frac{(1 - \Omega)\, r^2}{2c^2 \tau^2}. \tag{5.58}$$

When the contribution of the cosmological term r^2/a^2 is negligible, we then have

$$\frac{\lambda_2}{\lambda_1} = \frac{1}{\sqrt{1 - R_s/R}}. \tag{5.59}$$

The redshift could then be very large if R, the radius of the emitting body, is just a bit larger than the Schwarzchild radius R_s. For example if $R_s/R = 0.96$ the redshift is $z = 4$. For a typical sun like ours, $R_s \ll R$ and we can expand the righthand side of Eq. (5.59), getting

$$\frac{\lambda_2}{\lambda_1} = 1 + \frac{R_s}{2R}, \tag{5.60}$$

thus

$$z = \frac{R_s}{2R} = \frac{Gm}{c^2 R},\tag{5.61}$$

the standard general relativistic result.

From Eqs. (5.56)–(5.58) it is clear that Ω cannot be larger than one since otherwise z will be negative, which means blueshift, and as is well known nobody sees such a thing. If $\Omega = 1$ then $z = 0$, and for $\Omega < 1$ we have $z > 0$. The case of $\Omega = 1$ is also implausible since the light from stars we see is usually redshifted more than the redshift due to the gravity of the body emitting the radiation, as is evident from our sun, for example, whose emitted light is shifted by only $z = 2.12 \times 10^{-16}$.

5.5.3   Conclusions

One can thus conclude that the theory of cosmological general relavitity predicts that the Universe is open. As is well known the standard FRW model does not relate the cosmological redshift to the kind of Universe.

5.6   Verification of the classical general relativity tests

5.6.1   Comparison with general relativity

We first find the cosmological-generalization of the Schwarzschild spherically-symmetric metric in cosmology. It will be useful to change variables from the classical Schwarzschild metric to new variables as follows:

$$\sin^2 \chi = r_s/r, \quad dr = -2r_s \sin^{-3} \chi \cos \chi \, d\chi,\tag{5.62}$$

where $r_s = 2GM/c^2$ is the Schwarzschild radius. We also change the time coordinate $cdt = r_s d\eta$, thus η is a time parameter.

The classical Schwarzschild solution will thus have the following form:

$$ds^2 = r_s^2 \left[\cos^2 \chi d\eta^2 - 4\sin^{-6}\chi d\chi^2 - \sin^{-4}\chi \left(d\theta^2 + \sin^2\theta d\phi^2 \right) \right].$$
$$(5.63)$$

So far this is just the classical spherically symmetric solution of the Einstein field equations in four dimensions, though written in new variables. The non-zero Christoffel symbols are given by

$$\Gamma^0_{01} = -\sin\chi \cos^{-1}\chi, \qquad \Gamma^1_{00} = -\frac{1}{4}\sin^7\chi \cos\chi,$$

$$\Gamma^1_{11} = -3\sin^{-1}\chi \cos\chi, \qquad \Gamma^1_{22} = \frac{1}{2}\sin\chi \cos\chi,$$

$$\Gamma^1_{33} = \frac{1}{2}\sin\chi \cos\chi \sin^2\theta, \qquad \Gamma^2_{12} = -2\sin^{-1}\chi \cos\chi, \qquad (5.64)$$

$$\Gamma^2_{33} = -\sin\theta\cos\theta, \quad \Gamma^3_{13} = -2\sin^{-1}\chi \cos\chi, \quad \Gamma^3_{23} = \sin^{-1}\theta\cos\theta.$$

It is very lengthy, but one can verify that all components of the Ricci tensor $R_{\alpha\beta}$ are equal to zero identically.

We now extend this solution to cosmology. In order to conform with the standard notation, the zero component will be chosen as the time parameter, followed by the three space-like coordinates and then the fourth coordinate representing the velocity τdv. We will make one more change by choosing $\tau dv = r_s du$, thus u is the velocity parameter. The simplest way to have a cosmological solution of the Einstein field equation is using the so-called co-moving coordinates in which ds^2 is given by:

$$r_s^2 \left[\cos^2 \chi d\eta^2 - 4\sin^{-6}\chi d\chi^2 - \sin^{-4}\chi \left(d\theta^2 + \sin^2\theta d\phi^2 \right) + du^2 \right].$$
$$(5.65)$$

The coordinates are now $x^0 = \eta$, $x^1 = \chi$, $x^2 = \theta$, $x^3 = \phi$, and $x^4 = u$, and r_s is now a function of the velocity u, $r_s = r_s(u)$ to be determined by the Einstein field equations in five dimensions. Accordingly we have the following form for the metric:

$$g_{\mu\nu} = r_s^2 \begin{pmatrix} \cos^2\chi & & & & 0 \\ & -4\sin^{-6}\chi & & & \\ & & -\sin^{-4}\chi & & \\ & & & -\sin^{-4}\chi\sin^2\theta & \\ 0 & & & & 1 \end{pmatrix},$$

(5.66a)

$$\sqrt{-g} = 2r_s^5 \sin^{-7}\chi \cos\chi \sin\theta.$$

(5.66b)

The non-zero Christoffel symbols are given by

$$\Gamma_{01}^0 = -\sin\chi\cos^{-1}\chi, \quad \Gamma_{04}^0 = \dot{r}_s r_s^{-1},$$

$$\Gamma_{00}^1 = -\frac{1}{4}\sin^7\chi\cos\chi, \quad \Gamma_{11}^1 = -3\sin^{-1}\chi\cos\chi,$$

$$\Gamma_{14}^1 = \dot{r}_s r_s^{-1}, \quad \Gamma_{22}^1 = \frac{1}{2}\sin\chi\cos\chi, \quad \Gamma_{33}^1 = \frac{1}{2}\sin\chi\cos\chi\sin^2\theta,$$

$$\Gamma_{12}^2 = -2\sin^{-1}\chi\cos\chi, \quad \Gamma_{24}^2 = \dot{r}_s r_s^{-1}, \quad \Gamma_{33}^2 = -\sin\theta\cos\theta,$$

(5.67)

$$\Gamma_{13}^3 = -2\sin^{-1}\chi\cos\chi, \quad \Gamma_{23}^3 = \sin^{-1}\theta\cos\theta, \quad \Gamma_{34}^3 = \dot{r}_s r_s^{-1},$$

$$\Gamma_{00}^4 = -\dot{r}_s r_s^{-1}\cos^2\chi, \quad \Gamma_{11}^4 = 4\dot{r}_s r_s^{-1}\sin^{-6}\chi, \quad \Gamma_{22}^4 = \dot{r}_s r_s^{-1}\sin^{-4}\chi,$$

$$\Gamma_{33}^4 = \dot{r}_s r_s^{-1}\sin^{-4}\chi\sin^2\theta, \quad \Gamma_{44}^4 = \dot{r}_s r_s^{-1},$$

where the dots denote derivatives with respect to the velocity parameter u.

The Ricci tensor components after a lengthy but straightforward calculation, are given by:

$$R_{00} = -\left(\ddot{r}_s r_s^{-1} + 2\dot{r}_s^2 r_s^{-2}\right)\cos^2\chi,$$

$$R_{11} = 4\left(\ddot{r}_s r_s^{-1} + 2\dot{r}_s^2 r_s^{-2}\right)\sin^{-6}\chi,$$

$$R_{22} = \left(\ddot{r}_s r_s^{-1} + 2\dot{r}_s^2 r_s^{-2}\right)\sin^{-4}\chi,$$

(5.68)

$$R_{33} = \left(\ddot{r}_s r_s^{-1} + 2\dot{r}_s{}^2 r_s^{-2} \right) \sin^{-4} \chi \sin^2 \theta,$$

$$R_{44} = -4 \left(\ddot{r}_s r_s^{-1} - \dot{r}_s{}^2 r_s^{-2} \right).$$

All other components are identically zero.

We are interested in vacuum solution of the Einstein field equations for the spherically symmetric metric (Schwarzschild to cosmology), the right-hand sides of the above equations should be taken zero. A simple calculation then shows that $\dot{r}_s = 0$, $\ddot{r}_s = 0$. Accordingly the cosmological Schwarzschild metric is given by Eq. (5.66a) with a constant $r_s = 2GM/c^2$. The metric (5.66a) can then be written, using the coordinate transformations (5.62), as

$$g_{\mu\nu} = \begin{pmatrix} 1 - \frac{r_s}{r} & & & & 0 \\ & -\left(1 - \frac{r_s}{r}\right)^{-1} & & & \\ & & -r^2 & & \\ & & & -r^2 \sin^2 \theta & \\ 0 & & & & 1 \end{pmatrix}, \quad (5.69)$$

where the coordinates are now $x^0 = ct$, $x^1 = r$, $x^2 = \theta$, $x^3 = \phi$, and $x^4 = \tau v$.

We are now in a position to compare the present theory with general relativity. By that we mean the verification of the three classical tests of general relativity in cosmological general relativity. This will be done in almost identical way to that used for verifying these tests in general relativity given in Chapter 3, but now use has to be made by the cosmological Schwarzschild metric in five dimensions.

5.6.2 Gravitational redshift

We start with the simplest experiment, that of the gravitational redshift. This experiment is not considered as one of the proofs of general relativity (it can be derived from conservation laws and Newtonian theory).

Consider two clocks at rest at two points denoted by 1 and 2. The propagation of light is determined by ds at each point. Since at these points all spatial infinitesimal displacements and change in velocities vanish, one has $ds^2 = g_{00}c^2dt^2$. Hence at the two points we have

$$ds(1) = [g_{00}(1)]^{1/2} c\,dt, \tag{5.70a}$$

$$ds(2) = [g_{00}(2)]^{1/2} c\,dt \tag{5.70b}$$

for the proper time.

The ratio of the rates of similar clocks, located at different places in a gravitational field, is therefore given by

$$ds(2)/ds(1) = [g_{00}(2)/g_{00}(1)]^{1/2}. \tag{5.71}$$

The frequency ν_0 of an atom located at point 1, when measured by an observer located at point 2, is therefore given by

$$\nu = \nu_0 [g_{00}(1)/g_{00}(2)]^{1/2}. \tag{5.72}$$

If the gravitational field is produced by a spherically symmetric mass distribution, then we may use the generalized Schwarzschild metric to calculate the above ratio at the two points. In this case $g_{00} = 1 - 2GM/c^2r$, and therefore

$$[g_{00}(1)/g_{00}(2)]^{1/2} \approx 1 + \left(GM/c^2\right)(1/r_2 - 1/r_1)$$

to first order in GM/c^2r.

We thus obtain

$$\Delta\nu/\nu_0 = (\nu - \nu_0)/\nu_0 \approx -\left(GM/c^2\right)(1/r_1 - 1/r_2)$$

for the frequency shift per unit frequency. Taking now r_1 to be the observed radius of the Sun and r_2 the radius of the Earth's orbit around the Sun, then we find that

$$\Delta\nu/\nu_0 \approx -GM_{Sun}/c^2 r_{Sun}, \tag{5.73}$$

where M_{Sun} and r_{Sun} are the mass and radius of the Sun.

Accordingly we obtain $\Delta\nu/\nu_0 \approx -2.12 \times 10^{-6}$ for the frequency shift per unit frequency of the light emitted from the Sun. The calculation made above amounts to neglecting completely the Earth's gravitational field. The above result is the standard gravitational redshift (also known as the gravitational time dilation).

5.6.3 Motion in a centrally symmetric gravitational field

We assume that small test particles move along geodesics in the gravitational field. We also assume that planets have small masses as compared with the mass of the Sun, to the extent that they can be considered as test particles moving in the gravitational field of the Sun. As a result of these assumptions, the geodesic equation in the cosmological Schwarzschild field will be taken to describe the equation of motion of a planet moving in the gravitational field of the Sun. In fact, we do not need the exact solution of the cosmological Schwarzschild metric (5.69), but just its first approximation.

We obtain in the first approximation the following expressions for the components of the metric tensor:

$$g_{00} = 1 - r_s/r, \qquad g_{0m} = 0, \qquad g_{04} = 0,$$

$$g_{mn} = -\delta_{mn} - r_s x^m x^n/r^3, \qquad g_{m4} = 0, \qquad g_{44} = 1. \qquad (5.74a)$$

The contravariant components of the metric tensor are consequently given, in the same approximation, by

$$g^{00} = 1 + r_s/r, \qquad g^{0m} = 0, \qquad g^{04} = 0,$$

$$g^{mn} = -\delta^{mn} + r_s x^m x^n/r^3, \qquad g^{m4} = 0, \qquad g^{44} = 1. \qquad (5.74b)$$

We may indeed verify that the relation $g_{\mu\lambda}g^{\lambda\nu} = \delta^\nu_\mu$ between the contravariant and covariant components of the above approximate metric tensor is satisfied to orders of magnitude of the square of r_s/r. A straightforward calculation then gives the following expressions for the Christoffel symbols:

$$\Gamma^0_{0n} = -\frac{r_s}{2}\frac{\partial}{\partial x^n}\left(\frac{1}{r}\right),$$

$$\Gamma^k_{00} = --\frac{r_s}{2}\left(1-\frac{r_s}{r}\right)\frac{\partial}{\partial x^k}\left(\frac{1}{r}\right), \tag{5.75}$$

$$\Gamma^k_{mn} = r_s\frac{x^k}{r^3}\delta_{mn} - \frac{3}{2}r_s\frac{x^k x^m x^n}{r^5}.$$

All other components vanish.

We now use these expressions for the Christoffel symbols in the geodesic equation

$$\ddot{x}^k + \left(\Gamma^k_{\alpha\beta} - \Gamma^0_{\alpha\beta}\dot{x}^k\right)\dot{x}^\alpha\dot{x}^\beta = 0, \tag{5.76}$$

where a dot denotes differentiation with respect to the time coordinate x^0. We obtain

$$\Gamma^0_{\alpha\beta}\dot{x}^\alpha\dot{x}^\beta = \Gamma^0_{00}+2\Gamma^0_{0n}\dot{x}^n+2\Gamma^0_{04}\dot{x}^4+\Gamma^0_{mn}\dot{x}^m\dot{x}^n+2\Gamma^0_{m4}\dot{x}^m\dot{x}^4+\Gamma^0_{44}\dot{x}^4\dot{x}^4$$

$$= -r_s\dot{x}^n\frac{\partial}{\partial x^n}\left(\frac{1}{r}\right), \tag{5.77a}$$

$$\Gamma^k_{\alpha\beta}\dot{x}^\alpha\dot{x}^\beta = \Gamma^k_{00}+2\Gamma^k_{0l}\dot{x}^l+2\Gamma^k_{04}\dot{x}^4+\Gamma^k_{mn}\dot{x}^m\dot{x}^n+2\Gamma^k_{m4}\dot{x}^m\dot{x}^4+\Gamma^k_{44}\dot{x}^4\dot{x}^4$$

$$= -\frac{r_s}{2}\frac{\partial}{\partial x^k}\left(\frac{1}{r}\right)$$

$$+r_s\left[\frac{r_s}{2r}\frac{\partial}{\partial x^k}\left(\frac{1}{r}\right) - \left(\dot{x}^s\dot{x}^s\right)\frac{\partial}{\partial x^k}\left(\frac{1}{r}\right) - \frac{3}{2r^5}\left(x^s\dot{x}^s\right)^2 x^k\right]. \tag{5.77b}$$

Consequently we obtain from the geodesic equation (5.76) the following equation of motion for the planet:

$$\ddot{x}^k - \frac{r_s}{2} \frac{\partial}{\partial x^k} \left(\frac{1}{r} \right)$$

$$= r_s \left[\left(\dot{x}^s \dot{x}^s \right) - \frac{r_s}{2r} \right] \frac{\partial}{\partial x^k} \left(\frac{1}{r} \right)$$

$$- r_s \left[\dot{x}^n \frac{\partial}{\partial x^n} \left(\frac{1}{r} \right) \dot{x}^k - \frac{3}{2r^5} \left(x^s \dot{x}^s \right)^2 x^k \right]. \tag{5.78}$$

Replacing now the derivatives with respect to x^0 by those with respect to $t(\equiv x^0/c)$ in the latter equation, we obtain

$$\ddot{\mathbf{x}} - GM\nabla\frac{1}{r}$$

$$= r_s \left[(\dot{\mathbf{x}}^2)\nabla \left(\frac{1}{r} \right) - \frac{GM}{r} \nabla \left(\frac{1}{r} \right) - \left(\dot{\mathbf{x}} \cdot \nabla\frac{1}{r} \right) \dot{\mathbf{x}} + \frac{3}{2r^5} (\mathbf{x} \cdot \dot{\mathbf{x}})^2 \mathbf{x} \right], \tag{5.79}$$

where use has been made of the three-dimensional notation.

Hence the equation of motion of the planet differs from the Newtonian one since the left-hand side of Eq. (5.79) is proportional to terms of order of magnitude r_s instead of vanishing identically. This correction leads to a fundamental effect, namely, to a systematically secular change in the perihelion of the orbit of the planet.

To integrate the equation of motion (5.79) we multiply it vectorially by the radius vector $\mathbf{x}$. We obtain

$$\mathbf{x} \times \ddot{\mathbf{x}} = -r_s \left(\dot{\mathbf{x}} \cdot \nabla (1/r) \right) (\mathbf{x} \times \dot{\mathbf{x}}). \tag{5.80}$$

All other terms in Eq. (5.79) are proportional to the radius vector $\mathbf{x}$ and thus contribute nothing. Equation (5.80) may be integrated to yield the first integral

$$\mathbf{x} \times \dot{\mathbf{x}} = \mathbf{J}e^{-r_s/r}. \tag{5.81}$$

Here $\mathbf{J}$ is a constant vector, the *angular momentum* per mass unit of the planet. One can easily check that the first integral

(5.81) indeed leads back to Eq. (5.80) by taking the time derivatives of both sides of Eq. (5.81).

From Eq. (5.81) we see that the radius vector **x** moves in a plane perpendicular to the constant angular momentum vector **J**, thus the planet moves in a plane similar to the case in Newtonian mechanics. If we now introduce in this plane coordinates r and ϕ to describe the motion of the planet, the equation of motion (5.79) consequently decomposes into two equations. Introducing now the new variable $u = 1/r$, we can then rewrite the equations in terms of $u(\phi)$, using

$$\dot{r} = -\frac{u'}{u^2}\dot{\phi},$$

$$\ddot{r} = \frac{2u'^2}{u^3}\dot{\phi}^2 - \frac{u''}{u^2}\dot{\phi}^2 - \frac{u'}{u^2}\ddot{\phi},$$

where a prime denotes differentiation with respect to the angle ϕ. We subsequently obtain

$$\ddot{\phi} = 2\frac{u'}{u}\dot{\phi}^2 - \frac{2GM}{c^2}u'\dot{\phi}^2.$$

A straightforward calculation then gives, using the expression for $\ddot{\phi}$,

$$u'' + u - GM\left(\frac{u^2}{\dot{\phi}}\right)^2 = \frac{GM}{c^2}\left[2u^2 - u'^2 - 2GMu\left(\frac{u^2}{\dot{\phi}}\right)^2\right]$$

(5.82).

The latter equation can be further simplified if we use the first integral

$$r^2\dot{\phi} = Je^{-2GM/c^2r}.$$

We obtain

$$\frac{u^2}{\dot{\phi}} = \frac{1}{J}e^{2GMu/c^2},$$

$$\left(\frac{u^2}{\dot{\phi}}\right)^2 = \frac{1}{J^2}e^{4GMu/c^2} \approx \frac{1}{J^2}\left(1 + \frac{4GM}{c^2}u\right).$$

Hence, to an accuracy of $1/c^2$, Eq. (5.82) gives

$$u'' + u - \frac{GM}{J^2} = \frac{GM}{c^2}\left(2u^2 - u'^2 + 2\frac{GM}{J^2}u\right). \qquad (5.83)$$

Equation (5.83) can be used to determine the motion of the planet. The Newtonian equation of motion that corresponds to Eq. (5.83) is one whose left-hand side is identical to the above equation, but is equal to zero rather than to the terms on the right-hand side. This fact can easily be seen if one lets GM/c^2 go to zero in Eq. (5.83). Therefore in the Newtonian limit we have

$$u'' + u - \frac{GM}{J^2} \approx 0, \qquad (5.84)$$

whose solution can be written as

$$u \approx u_0\left(1 + \epsilon \cos \phi\right). \qquad (5.85)$$

Here u_0 is a constant, and ϵ is the eccentricity of the ellipse, $\epsilon = (1 - b^2/a^2)^{1/2}$, where a and b are the semimajor and semiminor axes of the ellipse. Using the solution (5.85) in the Newtonian limit of the equation of motion (5.84) then determines the value of the constant u_0, as $u_0 = GM/J^2$.

To solve the equation of motion (5.83), we therefore assume a solution of the form

$$u = u_0\left(1 + \epsilon \cos \alpha\phi\right), \qquad (5.86)$$

where α is some parameter to be determined, and whose value in the usual nonrelativistic mechanics is unity. The appearance of the parameter $\alpha \neq 1$ in our solution is an indication that the motion of the planet will no longer be a closed ellipse.

Using the above solution in Eq. (5.83), and equating coefficients of $\cos \alpha\phi$, then gives

$$\alpha^2 = 1 - \frac{2GM}{c^2}\left(2u_0 + \frac{GM}{J^2}\right).$$

If we substitute for GM/J^2 in the above equation its nonrelativistic value u_0, then the error will be of a higher order. Hence the latter equation can be written as

$$\alpha^2 = 1 - \frac{6GM}{c^2}u_0$$

or

$$\alpha = 1 - \frac{3GM}{c^2}u_0. \tag{5.87}$$

Successive perihelia occur at two angles ϕ_1 and ϕ_2 when $\alpha\phi_2 - \alpha\phi_1 = 2\pi$. Since the parameter α is smaller than unity, we have $\phi_2 - \phi_1 = 2\pi/\alpha > 2\pi$. Hence we can write $\phi_2 - \phi_1 = 2\pi + \Delta\phi$, with $\Delta\phi > 0$, or

$$\alpha\left(\phi_2 - \phi_1\right) = \alpha\left(2\pi + \Delta\phi\right) = \left(1 - \frac{3GM}{c^2}u_0\right)\left(2\pi + \Delta\phi\right) = 2\pi. \tag{5.88}$$

As a result there will be an *advance* in the perihelion of the orbit of the planet per revolution given by Eq. (5.88) or, to first order, by

$$\Delta\phi = 6\pi GMu_0/c^2. \tag{5.89}$$

The constant u_0 can also be expressed in terms of the eccentricity, using the Newtonian approximation. Denoting the radial distances of the orbit, which correspond to the angles $\phi_2 = 0$ and $\phi_1 = \pi$, by r_2 and r_1, respectively, we have from Eq. (5.85),

$$1/r_2 = u_0\left(1 + \epsilon\right), \quad 1/r_1 = u_0\left(1 - \epsilon\right).$$

Hence since $r_1 + r_2 = 2a$, we obtain

$$2a = r_1 + r_2 = 2/u_0\left(1 - \epsilon^2\right),$$

where a is the semimajor axis of the orbit, and therefore

$$u_0 = 1/a\left(1 - \epsilon^2\right).$$

Using this value for u_0 in the expression (5.89) for $\Delta\phi$, we obtain for the perihelion advance the expression

$$\Delta\phi = \frac{6\pi GM}{c^2 a\left(1-\epsilon^2\right)} \tag{5.90}$$

in radians per revolution. This is the standard general relativistic formula for the advance of the perihelion.

In the next subsection we discuss the deflection of a light ray moving in a gravitational field.

5.6.4 Deflection of light in a gravitational field

To discuss the effect of gravitation on the propagation of light signals we may use the geodesic equation, along with the null condition $ds = 0$ at a fixed velocity. A light signal propagating in the gravitational field of the Sun, for instance, will thus be described by the null geodesics in the cosmological Schwarzschild field at $dv = 0$.

Using the approximate solution for the cosmological Schwarzschild metric, given by Eq. (5.74a), we obtain

$$g_{\mu\nu}dx^\mu dx^\nu = \left(1 - \frac{2GM}{c^2 r}\right)c^2 dt^2 - \left[dx^s dx^s + \frac{2GM}{c^2}\frac{(x^s dx^s)^2}{r^3}\right] = 0. \tag{5.91}$$

Hence we have, to the first approximation in GM/c^2, the following equation of motion for the propagation of light in a gravitational field:

$$\left(1 + \frac{2GM}{c^2 r}\right)\left[\left(\dot{x}^s \dot{x}^s\right) + \frac{2GM}{c^2}\frac{\left(x^s \dot{x}^s\right)^2}{r^3}\right] = c^2, \tag{5.92}$$

where a dot denotes differentiation with respect to the time coordinate $t(\equiv x^0/c)$.

Just as in the case of planetary motion (see previous subsection), the motion here also takes place in a plane. Hence in this plane we may introduce the polar coordinates r and ϕ. The equation of motion (5.92) then yields, to the first approximation in GM/c^2, the following equation in the polar coordinates:

$$\left(\dot{r}^2 + r^2\dot{\phi}^2\right) + \frac{4GM}{c^2}\frac{\dot{r}^2}{r} + \frac{2GM}{c^2}r\dot{\phi}^2 = c^2. \tag{5.93}$$

Changing now variables from r to $u(\phi) \equiv 1/r$, we obtain

$$\left[u'^2 + u^2 + \frac{2GMu}{c^2}\left(2u'^2 + u^2\right)\right]\left(\frac{\dot{\phi}}{u^2}\right)^2 = c^2, \tag{5.94}$$

where a prime denotes differentiation with respect to the angle ϕ.

Moreover we may use the first integral of the motion,

$$r^2\dot{\phi} = Je^{-2GM/c^2r}, \tag{5.95}$$

in Eq. (5.94), thus getting

$$u'^2 + u^2 + \frac{2GMu}{c^2}\left(2u'^2 + u^2\right) = \left(\frac{c}{J}\right)^2 e^{4GMu/c^2}. \tag{5.96}$$

Differentiation of this equation with respect to ϕ then gives

$$u'' + u + \frac{GM}{c^2}\left(2u'^2 + 4uu'' + 3u^2\right) = \frac{2GM}{J^2}, \tag{5.97}$$

where terms have been kept to the first approximation in GM/c^2 only.

To solve Eq. (5.97) we notice that, in the lowest approximation, we have, from Eq. (5.96),

$$u'^2 \approx \left(\frac{c}{J}\right)^2 - u^2, \tag{5.98}$$

$$u'' \approx -u. \tag{5.99}$$

Hence using these approximate expressions in Eq. (5.97) gives

$$u'' + u = \frac{3GM}{c^2} u^2 \tag{5.100}$$

for the equation of motion of the orbit of the light ray propagating in a spherically symmetric gravitational field.

In the lowest approximation, namely, when the gravitational field of the central body is completely neglected, the right-hand side of Eq. (5.100) can be taken as zero, and therefore u satisfies the equation $u'' + u = 0$. The solution of this equation is a straight line given by

$$u = \frac{1}{R} \sin \phi, \tag{5.101}$$

where R is a constant. This equation for the straight line shows that $r \equiv 1/u$ has a minimum value R at the angle $\phi = \pi/2$. If we denote $y = r \sin \phi$, the straight line (5.101) can then be described by

$$y = r \sin \phi = R = \text{constant}. \tag{5.102}$$

We now use the approximate value for u, Eq. (5.101), in the right-hand side of Eq. (5.100), since the error introduced in doing so is of higher order. We therefore obtain the following for the equation of motion of the orbit of the light ray:

$$u'' + u = \frac{3GM}{c^2 R^2} \sin^2 \phi. \tag{5.103}$$

The solution of this equation is then given by

$$u = \frac{1}{R} \sin \phi + \frac{GM}{c^2 R^2} \left(1 + \cos^2 \phi\right). \tag{5.104}$$

Introducing now the Cartesian coordinates $x = r \cos \phi$ and $y = r \sin \phi$, the above solution can then be written as

$$y = R - \frac{GM}{c^2 R} \frac{2x^2 + y^2}{(x^2 + y^2)^{1/2}}. \tag{5.105}$$

We thus see that for large values of $|x|$ the above solution asymptotically approaches the following expression:

$$y \approx R - \frac{2GM}{c^2 R}|x|. \tag{5.106}$$

As seen from Eq. (5.106), asymptotically, the orbit of the light ray is described by two straight lines in the spacetime. These straight lines make angles with respect to the x axis given by $\tan \phi = \pm (2GM/c^2 R)$. The angle of deflection $\Delta\phi$ between the two asymptotes is therefore given by

$$\Delta\phi = \frac{4GM}{c^2 R}. \tag{5.107}$$

This is the angle of *deflection* of a light ray in passing through the gravitational field of a central body, described by the cosmological Schwarzschild metric. For a light ray just grazing the Sun, Eq. (5.107) gives the value

$$\Delta\phi = \frac{4GM_{\text{Sun}}}{c^2 R_{\text{Sun}}} = 1.75 \text{seconds}.$$

This is the standard general-relativistic formula. Observations indeed confirm this result. One of the latest measurements gives 1.75±0.10 seconds. It is worth mentioning that only general relativity theory and the present theory predict the correct factor of the deflection of light in the gravitational field.

5.7   Gravitational waves

In cosmological general relativity the theory also predicts a wave equation for gravitational radiation. In the linear approximation one obtains

$$\left(\frac{1}{c^2} \frac{\partial^2}{\partial t^2} - \nabla^2 + \frac{1}{\tau^2} \frac{\partial^2}{\partial v^2} \right) \gamma_{\mu\nu} = -2\kappa T_{\mu\nu}, \tag{5.108}$$

where $\gamma_{\mu\nu}$ is a first approximation term,

$$g_{\mu\nu} \approx \eta_{\mu\nu} + h_{\mu\nu} = \eta_{\mu\nu} + \gamma_{\mu\nu} - \eta_{\mu\nu}\gamma/2, \qquad (5.109)$$

$$\gamma = \eta^{\alpha\beta}\gamma_{\alpha\beta}. \qquad (5.110)$$

Hence CGR predicts that gravitational waves depend not only on space and time but also on the redshift of the emitting source.

References

S. Behar and M. Carmeli, *Intern. J. Theor. Phys.* **39**, 1375 (2000), astro-ph/0008352.

H. Bondi, Brandeis Summer School 1955.

M. Carmeli, *Phys. Rev.* **138**, B1003 (1965).

M. Carmeli, *Classical Fields: General Relativity and Gauge Theory* (John Wiley, New York, 1982).

M. Carmeli, *Cosmological Special Relativity: The Large-Scale Structure of Space, Time and Velocity*, Second Edition (World Scientific, River Edge, NJ. and Singapore, 2002).

M. Carmeli, *Intern. J. Theor. Phys.* **39**, 1397 (2000), astro-ph/9907244.

A. Einstein, *Autobiographical Notes*, Ed. P.A. Schilpp (Open Court Pub. Co., La Salle and Chicago, 1979).

V. Fock, *The Theory of Space, Time and Gravitation* (Pergamen Press, Oxford, 1959).

J.G. Hartnett and M.E. Tobar, Properties of gravitational waves in cosmological general relativity", *International Journal of Theoretical Physics*, in press, 2006, gr-qc/0603067.

P.J.E. Peebles, Status of the big bang cosmology, p. 84, in: *Texas/Pascos 92: Relativistic Astrophysics and Particle Cosmology*, Eds. C.W. Akerlof and M.A. Srednicki, Vol. 688 (The New York Academy of Sciences, New York, 1993).

B.C. Whitemore, Rotation curves of spiral galaxies in clusters, in: *Galactic Models*, J.R. Buchler, S.T. Gottesman, J.H. Hunter. Jr., Eds., (New York Academy Sciences, New York, 1990).

Appendix A

Mathematical Conventions

Throughout this appendix we use the convention

$$\alpha, \beta, \gamma, \delta, \cdots = 0, 1, 2, 3, 4,$$

$$a, b, c, d, \cdots = 0, 1, 2, 3,$$

$$p, q, r, s, \cdots = 1, 2, 3, 4,$$

$$k, l, m, n, \cdots = 1, 2, 3.$$

The coordinates are $x^0 = ct$, x^1, x^2 and x^3 are space-like coordinates, $r^2 = (x^1)^2 + (x^2)^2 + (x^3)^2$, and $x^4 = \tau v$. The signature is $(+ - - - +)$. The metric, approximated up to ϕ and ψ, is:

$$g_{\mu\nu} = \begin{pmatrix} 1+\phi & 0 & 0 & 0 & 0 \\ 0 & -1 & 0 & 0 & 0 \\ 0 & 0 & -1 & 0 & 0 \\ 0 & 0 & 0 & -1 & 0 \\ 0 & 0 & 0 & 0 & 1+\psi \end{pmatrix}, \qquad (A.1)$$

$$g^{\mu\nu} = \begin{pmatrix} 1-\phi & 0 & 0 & 0 & 0 \\ 0 & -1 & 0 & 0 & 0 \\ 0 & 0 & -1 & 0 & 0 \\ 0 & 0 & 0 & -1 & 0 \\ 0 & 0 & 0 & 0 & 1-\psi \end{pmatrix}. \qquad (A.2)$$

123

The nonvanishing Christoffel symbols are (in the linear approximation):

$$\Gamma^0_{0\lambda} = \frac{1}{2}\phi_{,\lambda}, \qquad \Gamma^0_{44} = -\frac{1}{2}\psi_{,0}, \qquad \Gamma^n_{00} = \frac{1}{2}\phi_{,n}, \tag{A.3a}$$

$$\Gamma^n_{44} = \frac{1}{2}\psi_{,n}, \qquad \Gamma^4_{00} = -\frac{1}{2}\phi_{,4}, \qquad \Gamma^4_{4\lambda} = \frac{1}{2}\psi_{,\lambda}, \tag{A.3b}$$

$$\Gamma^a_{00} = -\frac{1}{2}\eta^{ab}\phi_{,b}, \qquad \Gamma^a_{44} = -\frac{1}{2}\eta^{ab}\psi_{,b}, \tag{A.3c}$$

$$\Gamma^p_{00} = -\frac{1}{2}\eta^{pq}\phi_{,q}, \qquad \Gamma^p_{44} = -\frac{1}{2}\eta^{pq}\psi_{,q}. \tag{A.3d}$$

The Minkowskian metric η in five dimensions is given by

$$\eta = \begin{pmatrix} 1 & 0 & 0 & 0 & 0 \\ 0 & -1 & 0 & 0 & 0 \\ 0 & 0 & -1 & 0 & 0 \\ 0 & 0 & 0 & -1 & 0 \\ 0 & 0 & 0 & 0 & 1 \end{pmatrix}. \tag{A.4}$$

A.1 Components of the Ricci tensor

The elements of the Ricci tensor are:

$$R_{00} = \frac{1}{2}\left(\nabla^2\phi - \phi_{,44} - \psi_{,00}\right), \tag{A.5}$$

$$R_{0n} = -\frac{1}{2}\psi_{,0n}, \qquad R_{04} = 0, \tag{A.6}$$

$$R_{mn} = -\frac{1}{2}\left(\phi_{,mn} + \psi_{,mn}\right), \tag{A.7}$$

$$R_{4n} = -\frac{1}{2}\phi_{,4n}, \tag{A.8}$$

$$R_{44} = \frac{1}{2}\left(\nabla^2\psi - \phi_{,44} - \psi_{,00}\right). \tag{A.9}$$

The Ricci scalar is

$$R = \nabla^2\phi + \nabla^2\psi - \phi_{,44} - \psi_{,00}. \qquad \text{(A.10)}$$

The mixed Ricci tensor is given by

$$R_0^0 = \frac{1}{2}\left(\nabla^2\phi - \phi_{,44} - \psi_{,00}\right), \qquad \text{(A.11)}$$

$$R_0^n = \frac{1}{2}\psi_{,0n}, \qquad R_n^0 = -\frac{1}{2}\psi_{,0n}, \qquad \text{(A.12)}$$

$$R_0^4 = R_4^0 = 0, \qquad \text{(A.13)}$$

$$R_m^n = \frac{1}{2}\left(\phi_{,mn} + \psi_{,mn}\right), \qquad \text{(A.14)}$$

$$R_n^4 = -\frac{1}{2}\phi_{,n4}, \qquad R_4^n = \frac{1}{2}\phi_{,n4}, \qquad \text{(A.15)}$$

$$R_4^4 = \frac{1}{2}\left(\nabla^2\psi - \phi_{,44} - \psi_{,00}\right). \qquad \text{(A.16)}$$

APPENDIX B

INTEGRATION OF THE UNIVERSE EXPANSION EQUATION

The Universe expansion was shown to be given by Eq. (5.20),

$$\frac{dr}{dv} = \tau \left[1 + (1 - \Omega_M) r^2 / c^2 \tau^2 \right]^{1/2}.$$

This equation can be integrated exactly by the substitutions

$$\sin \chi = \alpha r / c\tau; \quad \Omega_M > 1 \tag{B.1a}$$

$$\sinh \chi = \beta r / c\tau; \quad \Omega_M < 1 \tag{B.1b}$$

where

$$\alpha = (\Omega_M - 1)^{1/2}, \quad \beta = (1 - \Omega_M)^{1/2}. \tag{B.2}$$

For the $\Omega_M > 1$ case a straightforward calculation using Eq. (B.1a) gives

$$dr = (c\tau / \alpha) \cos \chi d\chi \tag{B.3}$$

and the equation of the Universe expansion (5.20) yields

$$d\chi = (\alpha / c) \, dv. \tag{B.4a}$$

The integration of this equation gives

$$\chi = (\alpha / c) \, v + \text{const.} \tag{B.5a}$$

The constant can be determined using Eq. (B.1a). At $\chi = 0$, we have $r = 0$ and $v = 0$, thus

$$\chi = (\alpha/c)\, v, \tag{B.6a}$$

or, in terms of the distance, using (B.1a) again,

$$r(v) = (c\tau/\alpha)\sin \alpha v/c; \quad \alpha = (\Omega_M - 1)^{1/2}. \tag{B.7a}$$

This is obviously a decelerating expansion.

For $\Omega_M < 1$, using Eq. (B.1b), a similar calculation yields for the Universe expansion (5.30)

$$d\chi = (\beta/c)\, dv, \tag{B.4b}$$

thus

$$\chi = (\beta/c)\, v + \text{const.} \tag{B.5b}$$

Using the same initial conditions as above then gives

$$\chi = (\beta/c)\, v \tag{B.6b}$$

and in terms of distances,

$$r(v) = (c\tau/\beta)\sinh \beta v/c; \quad \beta = (1 - \Omega_M)^{1/2}. \tag{B.7b}$$

This is now an accelerating expansion.

For $\Omega_M = 1$ we have, from Eq. (5.20),

$$d^2 r/dv^2 = 0. \tag{B.4c}$$

The solution is, of course,

$$r(v) = \tau v. \tag{B.7c}$$

This is a constant expansion.

INDEX